Salman Shooshtarian
Tayyab Maqsood
Atiq Zaman
Savindi Caldera
Tim Ryley
and Peter SP Wong

Enhancing the Use of Products with Recycled Content in the Building and Construction Sector

nova
science publishers
www.novapublishers.com

https://doi.org/10.52305/KKEK0064

NOTICE TO THE READER

Library of Congress Cataloging-in-Publication Data

ISBN: 979-8-89113-634-2 (Hardcover)
ISBN: 979-8-89113-705-9 (eBook)

Published by Nova Science Publishers, Inc. † New York

Contents

List of Figures

List of Tables

Preface

With the continuous increase in construction and demolition (C&D) waste recovery rates in Australia, the next challenge is to establish a system for the sustainable application of PwRC in construction projects. In response to industry and public concerns related to unsustainable C&D waste management, this book aims to enhance understanding of the operation and capacity of end-markets for C&D waste recyclables in Australian jurisdictions. This understanding involves the integration of circular economy principles and a cradle-to-cradle approach into C&D waste management systems in the Australian context.

This publication is timely as it focuses on addressing the challenges of poor resource efficiency and inadequate performance in carbon emission mitigation in the architecture, engineering and construction (AEC) industry from a market development perspective. The resource efficiency is recognised as one of the six national research priorities. The book underscores the importance of incorporating circular economy principles in a comprehensive and cross-cutting manner to guide the management of C&D waste.

This book presents an extensive review of various national and international literature, an analysis of relevant policies and guidelines, and an evaluation of key stakeholders' sentiments. It provides valuable insights into five key areas: (1) analysis of national and jurisdictional policy frameworks that affect the creation and stimulation of end markets for waste materials; (2) assessment of the operation of existing end-markets for five strategic construction materials; (3) exploration of the application of sustainable procurement policies and practices at state, national and international levels; (4) evaluation of impacts of COVID-19 on the C&D waste industry; and (5) identification of key factors affecting C&D waste end-market creation and stimulation, and building a model for waste market development.

Collectively, the book aims to boost business prospects, provide advantages to product manufacturers, and establish sustainable circular supply chains.

The authors of this book are highly experienced academics with a well-established track record of conducting national research in the field of C&D waste management and the circular economy. Over the past six years, the authors have collaborated on three national research projects that have delved into various aspects of sustainable waste management within the Australian AEC industry. The findings and outcomes of this research have been widely disseminated through diverse communication channels, targeting industry, government, academic and public audiences. Additionally, the authors have actively engaged in policy advocacy efforts, making multiple submissions to local, jurisdictional, and national waste policies and strategies calls, further showcasing their commitment to driving positive change in the field of waste management.

Acknowledgments

This book is based on research that was developed by Australia's Sustainable Built Environment National Research Centre (SBEnrc). The SBEnrc develops projects informed by industry partner needs, secures national funding, project manages collaborative research and oversees research into practice initiatives. Current Core Members of the SBEnrc include ATCO Australia, BGC Australia, Government of Western Australia, Queensland Government, Sunshine Coast Council, Curtin University, Griffith University, Western Sydney University, and RMIT University. The industry-driven research described in this publication would not have been possible without the valuable support of our core industry, government and research partners.

We express our heartfelt gratitude to Ainsley Simpson, the CEO of the Infrastructure Sustainability Council, for her unwavering support in guiding the SBEnrc's Project (No. 1.85) Steering Group. Her expertise has greatly contributed to aligning the research direction with the industry's needs and concerns, ensuring a more effective and relevant outcome. We are also grateful for the editing support provided by Moein Behzad (School of Property, Construction and Project Management, RMIT University). The authors would also like to express their gratitude to David Baggs (LFRAIA), the CEO of Global GreenTag International, for his invaluable feedback and support throughout the planning and execution of our research.

Finally, the authors extend their sincere appreciation to the numerous individuals and organisations who have supported their research and the resulting book. These contributions have been invaluable and have greatly contributed to the successful completion of this project. Thank you for your support and commitment to advancing knowledge in this field.

List of Abbreviations

A&D	Architect and Design
AKDAR	Awareness, Desire, Knowledge, Action, Reinforcement
ABS	Australian Bureau of Statistics
ACT	Australian Capital Territory
ACCC	Australian Competition and Consumer Commission
ACOR	Australian Council of Recycling
AEC	Architecture, Engineering and Construction
BaU	Business as Usual
BE	Built Environment
Billion	Bn
BT	Billion tonnes
BRE	Building Research Establishment
BREAAM	Building Research Establishment Environmental Assessment Method
CAS	Chemical Abstracts Services
CDS	Container Deposit Scheme
C&D	Construction and Demolition
C&I	Commercial and Industrial
CE	Circular Economy
CRC	Crushed Recycled Concrete
CP	Client Participant
DfD	Design for Deconstruction
DfR	Design for Recycling
DfZw	Design for Zero-waste
D&C	Design and Construct
DGNB	German Sustainable Building Council
DoT	Department of Victoria
DT	Digital Technology
DTwins	Digital Twins

DCEEW	Department of Climate Change, Energy, the Environment and Water
DoW	Design out Waste
EfW	Energy from Waste
EoW	End of Waste
EoL	End of Life
EPA	Environmental Protection Authority
EMF	Ellen McArthur Foundation
EU	European Union
EPD	Environmental Product Declaration
GBCA	Green Building Council of Australia
GBCNZ	Green Building Council of New Zealand
GECA	Good Environmental Choice Australia
GDP	Gross Domestic Product
GHG	Greenhouse Gas Emissions
GISA	Green Industries SA
GPP	Green Public Procurement
GSP	Gross State Product
GFA	Gross Floor Area
HE	Higher Education
HDPE	High-density polyethene
IARC	International Agency for Research on Cancer
IEA	Institute of Engineers Australia
IPWEA	Institute of Public Works Engineering Australasia
ISC	Infrastructure Sustainability Council
ISO	International Organization for Standardisation
ILFI	International Living Future Institute
IPO	Input-process-output
Kt	Kilo tonne
KPI	key performance indicators
KWH	Kilowatt-hour
L3	Legacy Living Lab
LCA	Life Cycle Analysis
LCBA	Life Cycle Benefit Assessment
LEED	Leadership in Energy and Environmental Design
m	Million
MCDDJV	McConnell Dowell and Decmil Joint Venture
MCV	Manufacturer Claim Verification
MRF	Material Recycling Facility

MRPV	Major Road Projects Victoria
MRWA	MainRoad Western Australia
MFP	Mordialloc Freeway Project
MSW	Municipal Solid Waste
Mt	Million tonnes
NSW	New South Wales
NCC	National Construction Code
NC	No Consideration
NT	Northern Territory
NZ	New Zealand
NWP	National Waste Policy
OECD	Organisation for Economic Co-operation and Development
PwRC	Product with Recycled Content
PVC	Polyvinyl Chloride
PC	Partial Consideration
PCR	Product Category Rules
PRC	Recycled Product Certification
PP	polypropylene plastic
PRISMA	Preferred Reporting Items for Systematic Reviews and Meta-Analyses
PSG	Project Steering Group
PTV	Public Transport Victoria
PTA	Public Transport Authority
PHD	Product Health Declaration
Qld	Queensland
RAP	Reclaimed Asphalt Pavement
RPC	Recycled Product Certification
RtR	Roads to Reuse
Rd	Road
RMP	Rotationally Moulded Panel
SA	South Australia
GBCSA	Green Building Council of South Africa
SC	Significant Consideration
SDS	Safety Date Sheets
SESA	Skilling Educators for Sustainability Australia
SM	Salvaged Material
SMs	Sustainable Materials
SP	Sustainable Procurement
SPP	Sustainable Public Procurement

SV	Sustainability Victoria
SWOT	Strength, Weakness, Opportunity, Threat
t	Tonne (metric)
Tas	Tasmania
TAFE	Technical and Further Education
TfN	Transport for NSW
UA	Usable Area
UAE	United Arab Emirates
UDIA	Urban Development Institute of Australia
USGBC	U.S. Green Building Council
UN	United Nations
UK	United Kingdom
US	United States
USA	United States America
UV	Ultraviolet
UNEP	United Nations Environment Program
Vic	Victoria
VET	Vocational Education and Training
WA	Western Australia
WALGA	Western Australian Local Government Association
WMD	Waste Management Document
WMRR	Waste Management and Resource Recovery Association of Australia
WoS	Web of Science
WLCC	Whole Life-Cycle Cost
Yr	Year
3R	Reduction, Reuse, Recycling

Chapter 1

Introduction

1.1. Background

With rapid urbanisation around the world, construction activities continue to grow to an unprecedented extent. As a result, the building and construction sector now accounts for 35% of the total waste sent to landfills worldwide [1, 2]. For instance, in the UK, 49 Mt of construction and demolition (C&D) waste was generated in 2014, and more than half of this amount was landfilled [3]. In China, the annual C&D waste generation is 2.36 BT [1], with poor recovery rates [4], and in the US, this figure was 516 Mt in 2017 [5]. Despite the existing disparity in C&D waste recovery rates among OECD countries (Table 1), several estimates have predicted that fundamental changes are underway that will completely alter the C&D waste management environment. Recent analyses project that the global C&D waste recovery market will reach US $149–300 billion by 2027–28 [6, 7, 8].

Table 1. Latest rates of C&D waste generation and recovery in various countries

Country	Waste generated (Mt/)	Recovered (%)	Year	Source
Australia	29	77	2021	Blue Environment (2023) [9]
Brazil	48	20	2021	ABRELPE (2022) [10]
China	2,600	n/a	2020	Deng (2022) [4]
Germany	229	88	2020	Destatis (2021) [11]
India	150	1	2019	BMTPC (2021) [12]
Japan	n/a	97	2018	Ministry of Land Infrastructure Transport and Tourism (2020) [13]
The UK	67.8	92.3	2018	UK Government Statistical Service (2021) [14]
The US	600.3	76	2018	US EPA (2020) [15]
Russian Federation	15-17	5	2019	The Federal Service for State Statistics (2019) [16]
Italy	59	75.1	2018	ISPRA (2020) [17]
France	227	77	2020	ADEME (2021) [18]

The utilisation of products with recycled content (PwRC) and sustainable materials (SMs) has been suggested as a way to sustainably manage these resources. If planned properly, this management strategy will minimise waste disposal, keeping resource value in the economy loop as long as possible and reducing the need for new material extractions. Several theoretical initiatives such as the concepts of the 'circular economy' (CE), 'waste hierarchy model' and 'net-zero- (carbon) emissions' recommend using these resources. In addition, there is strong support from local, national and international waste policies to reuse PwRC in construction projects [19]. It is believed the use of PwRC can help governments meet their waste management targets [20]. Hence, the aim of this application is to address a variety of waste-related environmental, social and economic challenges.

On this basis, a few scientific enquiries in social science have explored the factors influencing the utilisation of these resources in construction projects. The outcomes have suggested that the utilisation of PwRC is not straightforward mainly because it is subject to a variety of internal and external factors that need to be well understood [21, 22]. Even the current utilisation of PwRC in the building and construction sector is primarily limited to low-value applications [23] otherwise known in waste management research as 'downcycling.' Previous research has provided reasons for these limitations such as energy and transport costs, lack of knowledge, limited technologies, poor quality and contamination, lack of market, unproductive waste regulations and negative perceptions [21]

1.2. Review of the Utilisation of PwRC in the Building and Construction Sector

This section of the chapter identifies the factors (enablers and barriers) that influence the use of PwRC in global construction projects. These factors were determined using a literature review. The research approach employed in this section is fully described in Chapter 2.

1.2.1. Barriers to Using PwRC in the Building and Construction Sector

The uptake of practices related to using PwRC is still limited and continues to have detrimental impacts on the environment [24]. The main barriers to using PwRC in the building and construction sector have been identified in previous studies (Table 2). The following section explains how these factors have hindered the application of PwRC in construction projects.

Table 2. Main barriers to the use of PwRC from stakeholders' perspectives

No	Barrier	Short description
1	Increased costs of energy and transport	PwRC are costlier than raw materials
2	Lack of knowledge of PwRC	Unfamiliarity with PwRC limited their application in construction projects
3	Limited technologies for waste recovery	Limited technologies in the waste recovery industry resulted in poor quality and expensive PwRC
4	Low quality, contamination and reduced performance	Lower than expected quality hinders the wide application of PwRC in the industry
5	Lack of market availability of the products	Uncertainty about the market discourages the production of quality second-hand materials
6	Limitations caused by specifications, standards and permits	Specifications and standards have both positive and negative impacts on the PwRC market
7	Limited acceptability and negative perceptions (public and the industry)	Negative perceptions toward PwRC limited their usage in the industry

1.2.1.1. Increase in Energy and Transport Costs

Cost is considered to be one of the main barriers to using PwRC in the building and construction sector (Figure 1) [25]. Several studies indicate that PwRC are costlier than conventional materials [26, 27, 28, 29]. A few reasons that contribute to higher costs in PwRC include longer transport distance [30], industrial waste sorting procedures [26], increased energy requirements for recycling, landfill levy for residual waste and high capital costs for recycling facilities [31, 32] and equipment [32, 33]. In some circumstances, the use of PwRC is reported to be less expensive than fresh materials [34, 35]. However, if waste producers and recyclers are willing to work collaboratively, it can lead to a win-win situation for both parties [35].

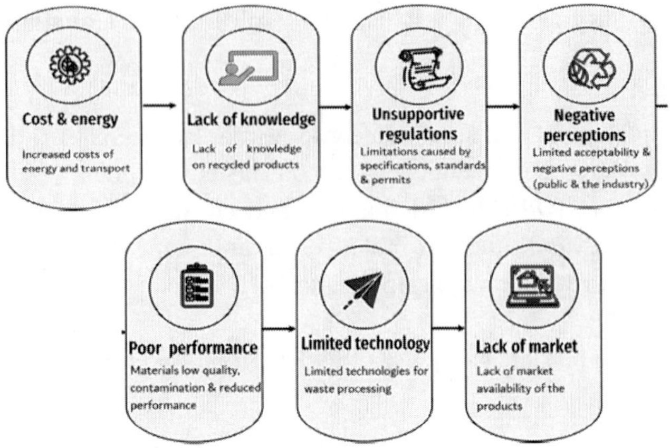

Figure 1. The primary barriers to using PwRC in construction projects.

1.2.1.2. Limited Knowledge of PwRC

There is solid evidence that a lack of knowledge of the various characteristics of recycled waste is a factor that diminishes the willingness to use them in the building and construction sector. A lack of awareness and knowledge leads to negative market demand for PwRC. This could also create a negative perception of used materials, final product quality and price in PwRC from end-users [36, 37]. Marketplaces for re-used and PwRC are yet to become mainstream, and there is still limited development of knowledge and practical impacts [26]. The limited communication between key stakeholders such as contractors and recyclers and a 'wait and see' attitude are other factors also affecting the limited knowledge of PwRC [26].

1.2.1.3. Limited Technologies for Waste Recovery

Technologies play an important role in safe and effective waste recovery. Limited technological advances and associated high capital costs in the waste recovery sector historically hinder the manufacturing of commercially competitive products [37, 38]. Limited education on emerging waste recovery technologies, lack of investment in waste management technologies [26] and limited incentives for private investment in technology [39] are key reasons for the poor uptake of PwRC use [22].

1.2.1.4. Low Quality and Reduced Performance

Previous research studies indicate that building and construction sector stakeholders face issues such as quality, contamination and unacceptable performance when choosing PwRC over conventional materials [36, 37]. Lack of technological support, such as resources, training [26], limited competent staff and expertise, and the lack of a benchmarking tool [26], plus the lack of systematic decontamination procedures [40] are key reasons affecting the quality and reduced performance of PwRC. As there is no assurance of maintaining the same quality and performance as virgin resources in the PwRC, there is a critical need for defining standards, producers and training staff members [26].

1.2.1.5. Lack of Market Availability of the Products

The viable market has a direct mutual relationship with further using PwRC. When there is an established market for PwRC, the economy of scale assists with further investment in recycling facilities, resulting in lower total cost, higher quality and less contamination. The imbalance of supply and demand for PwRC creates a narrow window of opportunity to trade recycled C&D products [31]. A study in Australia on recycled PVC products application reports that the cost of PwRC is the same as virgin material and recommended that the industry needs greater scale or a higher value for PVC to incentivise more PVC recycling [36]. Once these results are achieved, more industry stakeholders will be persuaded to buy and use PwRC in construction projects. Market availability is underpinned by several factors including an effective supply chain, properly designed landfill levy, government incentives and subsidies, and finding new applications outside of the building and construction sector [26].

1.2.1.6. Limitations Caused by Specifications, Standards and Permits

There is an ongoing debate within the waste management and resource recovery industry about the role of specifications, standards and permits in promoting the application of PwRC in the building and construction sector [20, 41]. If the organisations need to provide indemnities, operate trials or accept risk, these also create an additional layer of complexity that may become impediments to recycling C&D waste products [26]. On the one hand, specifications and standards determine the quality of materials recycled, giving peace of mind to stakeholders to use these materials. On the other hand, many recyclers indicate that complicated and difficult-to-meet specifications and standards are a significant hindrance in selling PwRC to the market.

1.2.1.7. Limited Acceptability and Negative Perceptions (Public and the Industry)

Due to various factors, such as a lack of knowledge and a view of the products as unconventional, there is a negative perception among the sector stakeholders towards using PwRC [26]. Different stakeholder groups may have different perceptions [42]. These perceptions are changing as field trials show how to use these materials to their optimum performance and as virgin resources become scarcer [36]. This highlights the critical need for prototyping during design stages to test acceptability [26].

1.2.2. Enablers to using PwRC in the Building and Construction Sector

Five clusters of enablers have been identified that can assist key stakeholders in advocating the use of PwRC in the sector (Figure 2).

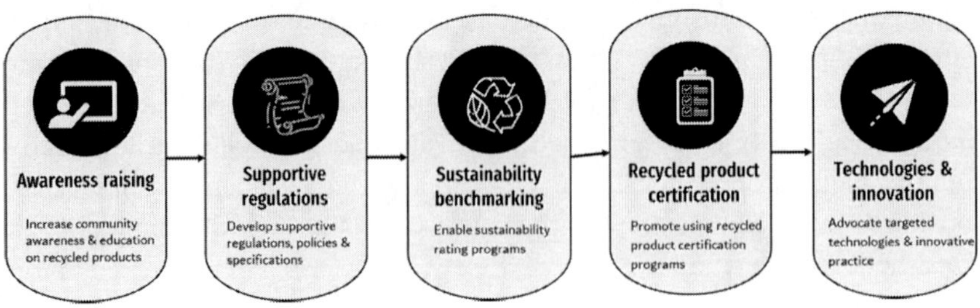

Figure 2. Main enablers of using PwRC in construction projects.

1.2.2.1. Community Awareness and Education on PwRC

Access to information and the education of construction experts are the two points of leverage for increasing the application of PwRC [43]. Therefore, bolstering efforts to inform stakeholders about the technical properties and environmental performance (primarily awarding authorities) and existing regulations and standards, is required. Furthermore, the distribution of more data, better documentation and reports about reference buildings assist with increasing engineers' experience with PwRC. In addition, the scientific community needs to make a more significant effort to convey the acquired knowledge to professionals.

1.2.2.2. Supportive Regulations, Policies and Specifications

The government should prioritise the task of improving specific legislation and regulations, with a focus on establishing a mandatory degree of normative standards. Notably, design codes should highlight the environmental aspects of the construction and give designers more flexibility in material sourcing leading to an increased usage of PwRC [21]. Detailed specifications should also be developed in the form of application-specific quality standards for different PwRC as another option to increase the demand for such products [43]. The introduction of mandatory policies (unified, stakeholder-oriented policies) [26, 44], such as sustainable procurement to meet specific reuse /recycling targets, more purchasing power and grants that are offered to companies to help initiate recycling locally [45] would encourage both recyclers and builders to invest in waste management solutions [37, 45].

1.2.2.3. Enabling Sustainability Programs

Another promising route is the implementation of sustainability programs that suggest the use of PwRC in construction projects. Some reports [46, 47] have demonstrated that running these programs leads to further adoption of PwRC in the building and construction sector. However, there is also evidence proving otherwise [48]. Among others, rating system design, developers' biases and lack of incentives to motivate end-users are the most influential reasons cited for these programs' ineffectiveness. Furthermore, scholarly and political engagement with business practice is urgently needed to facilitate sustainability programs [49].

1.2.2.4. Recycled Product Certification (PCR)

A product certification that is awarded following material testing, plus quality control, can increase the adoption of PwRC in the building and construction sector. As different supply sources will cause variation in quality, stricter quality controls for PwRC are required [26]. A government agency should be established to control the quality and production of these materials. Guidance on clear technical specifications or standards on the use of recycled aggregate for structural applications will enable recyclers to produce and adhere to industry standards [41].

1.2.2.5. Targeted Technologies and Innovative Practice

The waste management and resource recovery industry should embrace new technologies and innovation. Technological advances help PwRC producers meet market demands through products having reasonable prices and quality.

Notably, governments, as the main stakeholders responsible for protecting the environment, should support industry by providing a fund to invest in technologies and innovation.

1.3. Conclusion

This chapter provided an overview of the main challenges and drivers of using PwRC in the building and construction sector through the analysis of the existing literature. The analyses showed that the main enablers of using PwRC in the construction projects include (1) community awareness and education on PwRC, (2) supportive regulations, policies and specifications, (3) enabling sustainability programs, (4) recycled product certification and (5) targeted technologies and innovative practice. These enablers can mitigate the impacts of some of the barriers to using PwRC in construction projects. These barriers include (1) cost and energy, (2) lack of knowledge, (3) unsupportive regulations, (4) negative perceptions, (5) PwRC's poor performance, (6) limited technology, and (7) lack of market.

1.3.1. Book Structure

This book consists of 12 chapters, outlined in Figure 3. In this introductory chapter, a comprehensive literature review indicates the primary enablers and barriers associated with the global implementation of PwRC in construction projects (Chapter 1). Subsequently, Chapter 2 describes the research approach employed to generate the insights expounded upon in the ensuing chapters (Chapters 3-11). Chapters 3 through 6 immerse readers in an analysis of four construction projects wherein PwRC were utilised. These case studies assess the motivations driving their adoption, barriers encountered, and the strategic approaches to be employed for optimal adoption of these resources within the Australian building and construction sector.

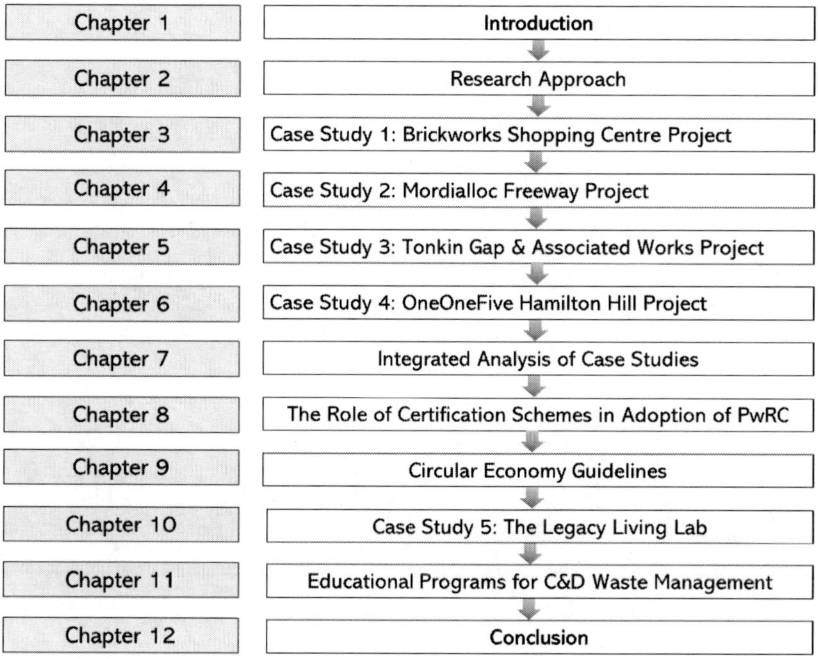

Figure 3. Structure of the book.

Chapter 7 presents the integrated analysis findings derived from the case study information. In Chapter 8, the efficacy of recycled product certification schemes in fostering PwRC utilisation is explored alongside stakeholder opinions. Chapter 9 examines circular economy guidelines, emphasizing circular design in construction. This is followed by a case study in Australia assessing guideline application (Chapter 10). Chapter 11 emphasises educating stakeholders on sustainable C&D waste management, sharing insights from training in two Australian organisations. Finally, Chapter 12 serves as the culminating piece, offering a brief synthesis that holds the key highlights and crucial takeaways from the preceding chapters. This concluding chapter provides a comprehensive wrap-up, elucidating the overarching themes and insights elucidated throughout this scholarly exploration.

Chapter 2

Research Approach

2.1. Introduction

This chapter presents the research approach and methods employed to produce information that shapes this book's content. The methods described in this chapter were used between 2022 and 2023 to achieve the following primary objectives:

1. To identify the main enablers of and barriers to using PwRC
2. To analyse the utilisation of these products in four Australian construction projects
3. To explore industry stakeholders' perceptions of using PwRC
4. To develop circular design guidelines in the building and construction sector
5. To evaluate the effectiveness of the pilot training sessions on C&D waste management

2.2. Research Methods

The book content is based on the findings that emerged from the application of various research methods (Figure 4), representing the use of a mixed-methods strategy to achieve the study objectives outlined above. The purpose of the mixed method was to collect both primary and secondary data. As presented in Figure 4, the research process consists of five stages, beginning with a literature review and culminating in some recommendations. This study's three data collection methods included a literature review (Chapters 1, 8-11), a case study analysis (Chapters 3-6), and a post-training survey (Chapter 11). The following section describes the research approach employed in this study.

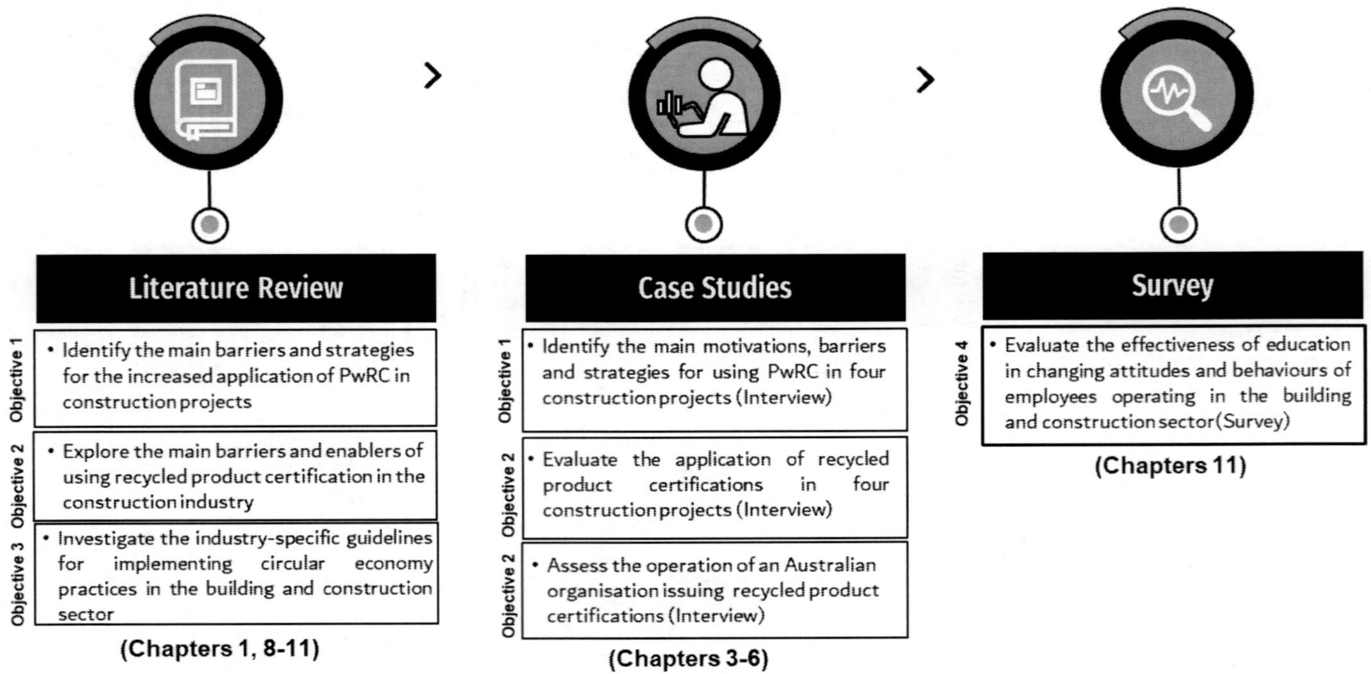

Figure 4. The main research methods employed.

2.3. Literature Review

This method was used in five chapters as outlined in Figure 4. In Chapter 1, an extensive literature review was carried out to identify the factors known to impact the use of PwRC in construction projects in Australia. The sources used in this step included journal articles and government and industry reports that are freely available online. Care was taken to only capture the literature that describes the use of PwRC in recent years to reflect the contemporary efforts and advances in using these materials.

In Chapter 8, the focus of the literature review was on identifying the primary obstacles and enablers of recycled product certifications (Chapter 8). The analysis was conducted in two parts, the first of which summarises the research findings pertaining to the Australian experience with such programmes, followed by an examination of the worldwide environment. Systematically, data were taken from academic papers, business and government reports, policies and guidelines, and reports produced by industry associations and other organisations.

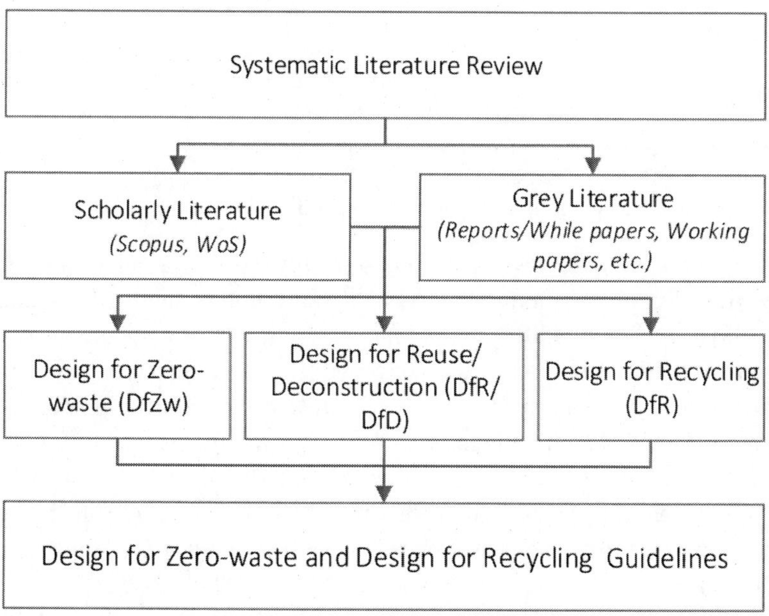

Figure 5. The research framework and scope in Chapter 9.

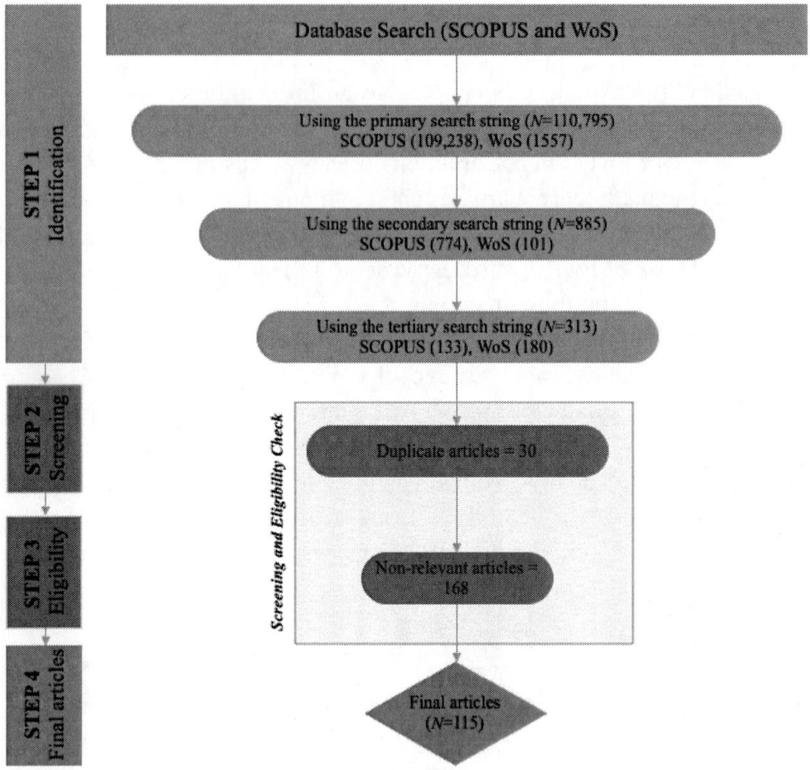

Source: Adopted from Moher et al., (2009) [50].

Figure 6. PRISMA diagram showing the selection process of the articles.

In Chapter 9, the systematic literature review was applied through the Preferred Reporting Items for Systematic Reviews and Meta-Analyses (PRISMA) technique [50]. This technique was used to investigate the circular design principles and associated concepts. The systematic literature review considered two scholarly databases (SCOPUS and Web of Science-WoS) using different search strings. Figure 5, shows the research framework employed in Chapter 9.

Only journal articles written in English were considered in the search criteria. For the grey literature, available reports, white papers and working papers from the relevant governments and industry bodies (e.g., Green Building Advisor, Australian Institute of Building, etc.) were considered. Three-level (primary, secondary and tertiary) search string criteria were used to identify relevant articles from more generic to specific topic areas. Table 3 shows the search strings used in different search criteria. At the end of five search attempts, 313 articles were preliminarily identified from both databases.

The three-level search strings help identify the relevant articles for construction and infrastructure projects. After identifying the relevant articles, screening and eligibility criteria were applied to finalise the articles for the meta-analysis. This stage resulted in the exclusion of 198 articles and finally, 115 articles were identified for the meta-analysis. Figure 6 shows the article selection steps through the PRISMA technique.

Following the analysis of relevant literature, the findings were used to benchmark the circular performance of a (construction) case study constructed in Perth, Western Australia (Chapter 10). This project is known as Legacy Living Lab (L3), implemented by Curtin University, and is a pioneering example of DfD and the circular economy in action (Figure 7). The project aims for a versatile, adaptable space that can be efficiently deconstructed and reassembled.

Source: L3 Lab, Curtin University [51].

Figure 7. The process of building a sustainable and modular building.

In Chapter 11, the research delved into an extensive literature review methodology to examine the role of education within the C&D waste management system, specifically in the context of Australia. Moreover, in this chapter, the relevant literature was reviewed to explore assessing frameworks that facilitate the evaluation of training course quality by establishing key metrics required for such assessments. The keywords used for extracting relevant literature included 'Australia,' 'education,' 'training,' 'construction and demolition waste,' and 'circular economy. The main search engines used in this study were Google and Google Scholar as their outputs consist of both academic and non-academic literature.

2.4. Case Study

The case study approach was utilised to explore the experience of the Australian key stakeholders in using PwRC in their projects (Chapters 3-6). In consultation with the waste experts, the criteria for the selection of case studies were set. Case studies had to meet the following criteria:

1. Significant use of recycled and/or processed PwRC; and
2. Completed recently or due to be completed.

Source(s): (in order of appearance from top left to bottom right): Armao (2021) [53], MainRoads Western Australia (2022) [54], TABEC (2022) [55], Victoria Big Build (2021) [56].

Figure 8. Images of case studies. Top left: Brickworks Shopping Centre (commercial/Case Study; 1); Top right: Mordialloc Freeway (infrastructure/Case Study 2); Bottom left: Tonkin Gap Highway (infrastructure/ Case Study 3); Bottom right: OneOneFive Hamilton Hill (residential/Case Study 4).

The selected case studies included two infrastructure (road transport) projects, one commercial project and one residential project. Except for one project (Burwood Brickworks Shopping Centre), all projects were government-owned projects [52]. The projects were built in two Australian states, Western Australia and Victoria, between 2018 and 2022 with budgets ranging from $2.7 M AUD to $400 M AUD. As shown in Table 4, the selected projects used significant quantities of PwRC in their construction activities. These projects have been recognised for outstanding levels of commitment to sustainability through different national sustainability recognition programs such as the Infrastructure Sustainability Council (ISC), Green Building Council of Australia (GBCA) and Urban Development Institute of Australia (UDIA). The projects have been referred to as demonstration projects for circular economy implementation on national and international levels (Figure 8).

After identifying the projects, the perceptions of four stakeholder groups playing an essential role in using PwRC were sought (Chapters 3-6). These groups included the designer, client, head-contractor and supplier (Figure 9). An invitation email was sent to each stakeholder to arrange suitable interview times. For each project selected, four separate online interviews were organised. The interviews were conducted online through Microsoft Teams between April 2022 and January 2023. In addition to questions related to the participants' demographic details, working experience and relevant expertise in the use of PwRC in construction projects, interviewees were asked about the main motivations, challenges and opportunities for the increased application of these materials in the selected case studies. Overall, 16 interviews were conducted before theoretical saturation was reached.

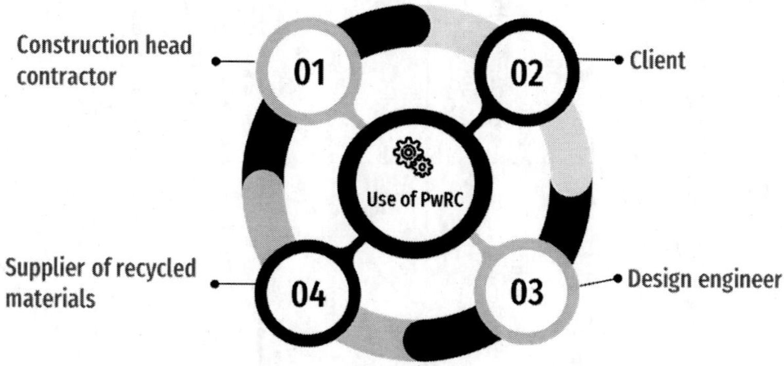

Figure 9. Stakeholders who were interviewed in each case study.

Table 3. Search strings used in the PRISMA approach

Step	Primary search criteria	Secondary search criteria	Tertiary search criteria	No available articles in the primary search		No available articles in the secondary search		No available articles in the tertiary search	
				Scopus	WoS	Scopus	WoS	Scopus	WoS
1	Design for 'zero-waste' OR design for 'no waste'	Design for 'zero-waste' OR design for 'no waste' AND construction OR infrastructure project	(Design for 'zero-waste' OR design for 'no waste') AND (construction OR infrastructure project) AND (guidelines OR strategies OR policies OR procedures OR standards OR protocols OR framework)	4228	384	89	20	43	38
2	Design for reuse OR design for recovery	Design for reuse OR design for recovery OR design for deconstruction AND construction OR infrastructure project	(Design for reuse OR design for recovery OR design for deconstruction) AND (construction OR infrastructure project) AND (guidelines OR strategies OR policies OR procedures OR standards OR protocols OR framework)	73361	48	34	4	20	29
3	Design for recycling OR design for resource recovery	Design for recycling OR design for resource recovery AND construction OR infrastructure project	(Design for recycling OR design for resource recovery) AND (construction OR infrastructure project) AND (guidelines OR strategies OR policies OR procedures OR standards OR protocols OR framework)	3726	190	80	7	44	39

Step	Primary search criteria	Secondary search criteria	Tertiary search criteria	No available articles in the primary search		No available articles in the secondary search		No available articles in the tertiary search	
				Scopus	WoS	Scopus	WoS	Scopus	WoS
4	Design for 'zero-waste' OR design for 'no waste' OR design for CE OR circular design	Design for 'zero-waste' OR design for 'no waste' OR design for CE OR circular design AND construction OR infrastructure project	(Design for 'zero-waste' OR design for 'no waste' OR design for CE OR circular design) AND (construction OR infrastructure project) AND (guidelines OR strategies OR policies OR procedures OR standards OR protocols OR framework)	2324	603	12	35	7	48
5	Design for deconstruction OR design for disassembly	Design for deconstruction OR design for disassembly AND construction OR infrastructure project	(Design for deconstruction OR design for disassembly AND (construction OR infrastructure project) AND (guidelines OR strategies OR policies OR procedures OR standards OR protocols OR framework)	25599	332	569	35	19	26
Total Identified articles from SCOPUS and WoS databases in different search levels				109238	1557	784	101	133	180
Total articles excluded due to eligibility check (duplication, non-relevant, etc.)				198					
The selected articles for review (N)				115					

Table 4. Overview of the selected case studies

Project features	Case Study 1	Case Study 2	Case Study 3	Case Study 4
Name of project	Burwood Brickworks Shopping Centre	Mordialloc Freeway	Tonkin Gap and Associated Works	OneOneFive Hamilton Hill
Construction type	Commercial	Infrastructure	Infrastructure	Residential
Client	Frasers Property Group	Major Road Projects Victoria	MainRoads Western Australia	WA Development
Budget (AUD m)	$120 m	$375 m	$400 m	$2.7 m
Demonstration of sustainability	• Living Building Challenge® [57] • Green Star (GBCA) [58]	• ISC's rating tool V.2 [59]	• ISC's rating tool V.2	• UDIA [60]
Project duration	2018–2019	2019–2021	2021–2023	2018–2019
Recycled products used (quantity/tonnes where provided)	• The use of crushed concrete in the sub-base of bitumen • The use of materials from slab form working as hanging timber and timber cladding in the ceiling	• The use of 600 t of plastic waste in noise walls • The use of 270 kt of pavement material incorporating the maximum allowable recycled content	• The reuse of 296 kt of sand • The reuse of 105 kt of spoil (treated ASS)	• The use of salvaged timber in landscaping features (e.g., shade structures and seating) • The reuse of 40,000 clay bricks and roof tiles as aggregates underneath the drainage infrastructure • The reuse of old bricks to create brick walls and a brick toilet block

Project features	Case Study 1	Case Study 2	Case Study 3	Case Study 4
Project duration	2018–2019	2019–2021	2021–2023	2018–2019
	• The use of second-hand brick purposed into tiles and concrete in floors • The use of crushed brick leftover as a finish on facades	• The use of 30 t of plastic waste in 100% recycled polypropylene plastic (PP) concrete reinforcing mesh • The use of 75 t of plastic waste in 100% recycled high-density polyethene (HDPE) stormwater drainage pipe	• The use of 27 kt of crushed recycled concrete (CRC) • The use of 1.2 kt of reclaimed asphalt pavement	• The reuse of crushed brick, tiles, concrete, etc., into the road sub-base replacing the need for newly mined materials like crushed limestone • The use of 2,425 m3 of recycled concrete in retaining walls replacing the need for newly mined limestone • The use of 400 t of PwRC in different applications including constructing temporary truck access roads
Product specifications	Different building codes and standards	407 VicRoads Standard Specifications: Dense Graded Asphalt [61]	Roads to Reuse Product Specifications [62]	IPWEA/WALGA Specification for the Supply of Recycled Road Base [63]

Audio data captured from the interviews were carefully transcribed by a professional transcriber word-for-word prior to the quality verification of text data by the research team. The transcripts were analysed using the NVivo Pro 12 application, which facilitates codifying text-based qualitative data. A deductive (theory-driven) coding system was initially adopted, using NVivo software, while new codes were created inductively from the interview data. A thematic analysis was applied to identify the emerging themes that were related to the three research objectives. Three coders were involved in the thematic analysis to achieve 80% inter-coder reliability. Furthermore, to better understand the qualitative data, a descriptive analysis of the data was undertaken. The main approach for these comparisons was frequency distributions of various categories of identified themes in the interviews. The constant comparative method (whereby interpretations and emergent findings are compared with existing findings) was employed to corroborate the interview findings.

Similar processes were used to analyse the experience and perspectives of industry stakeholders regarding the use of recycled product certification (Chapter 8). The interviewees were questioned about their opinions regarding the application of certification schemes in the case projects. The questions that were asked during interview times were as follows:

- Did your organisation consider recycled product certification during planning for this project? If so, what it was?
- What is your opinion about product certifications for recycled products? Do you know of any product certifiers in Australia or overseas? What are their limitations?

In Chapter 8, the operation of an existing RPC issuing organisation was also investigated. The studied organisation was based in Queensland, Australia servicing the global market [64]. A comprehensive description of this organisation is provided in Chapter 8. This case study involved interviewing a member of this organisation that took place in August 2022. The following questions were posed during the interview:

- What is the process of certifying PwRC in your organisation?
- Is this process aligned with state road product specifications and sustainability rating schemes?
- What is the mechanism of quoting/pricing for your organisation certification?

- How do you perceive the use of recycled product certification in today's built environment climate?
- In your opinion, what are the main reasons why businesses do not adopt recycled product certifications and what are the best strategies to promote using them in the built environment sector?
- What are the main benefits of using recycled product certification?
- Who are the main stakeholders that influence using the certifications?

2.5. Post-Training Evaluation through a Survey

Chapter 11 delves into the impacts of education on sustainable C&D waste management and assesses the effectiveness of two pilot training modules conducted for Australian public organisations in Western Australia, offering a deeper insight into their outcomes and implications. The former part was done through a questionnaire survey administered to understand how the stakeholders perceive the effectiveness of the training programs delivered to them.

The educational training module was aimed at individuals involved in the management and decision-making for solid waste systems which recycle and dispose of waste generated as a result of the construction, deconstruction, renovation and demolition of buildings and structures. The intended audience included design and structural engineers, project managers and construction managers. The module was pilot-tested with two Western Australia-based government organisations from the infrastructure sector and the building sector. The training effectiveness was evaluated through a survey inspired by the Kirkpatrick model [65]. Intended outcomes of the training include:

1. Identify environmental, economic and health impacts of C&D waste;
2. Define circular economy;
3. Describing the types of material and market opportunities and
4. learn about opportunities for creating a circular economy.

The content of the training packages was informed by the author's longitudinal study on construction and demolition management. The training materials included content related to unpacking the waste issue, circular economy principles, regulations, materials and market opportunities. Two training packages were developed for two Western Australia-based government organisations from the infrastructure sector and the building sector. These packages provide the knowledge required to manage C&D waste operations and develop environmental and identify opportunities for creating a circular economy. Understanding how to manage C&D waste streams the recycling potential and enhancing the uptake of using products with recycling materials is critical to mainstream sustainable waste management practices.

Stakeholder insights were gathered throughout the study, regarding their appreciation of the existing context, priorities, opportunities and challenges related to C&D waste management and creating a circular economy. Discussions were undertaken to explore stakeholder perspectives on their expectations for the educational and training program, following Griffith University Human Research ethics approval requirements (GU Ref No: 2022/153). Initially, a draft of the training package content was presented to the project steering group (PSG) members at the PSG meeting in October 2022. All feedback was taken onboard to revise the content. Specific feedback was provided on unpacking the C&D waste problem, statistical evidence (international and local data), categories of problems, nuances of circular economy definition, material examples and principles of the circular economy.

Then two in-depth face-to-face consultations were carried out with industry partner organisations between 1-3 December 2022. Considering the feedback and insights from the industry partners an educational and training program was co-developed for both the infrastructure and building sectors. The modules were delivered through either a one-hour or one-and-a-half-hour online session.

Chapter 3

Brickworks Shopping Centre Project (Case Study 1)

3.1. Project Background

Burwood Brickworks Shopping Centre was completed in 2019 in an eastern suburb of Melbourne, Victoria. The $120 m project is owned by Frasers Property Australia and was constructed by HACER Group. As a dining and entertainment precinct, this shopping centre provides shopping, dining and entertainment opportunities for Victorians (Figure 10).

Sources: Joe Armao & shoppingcentrenews.com.au [53].

Figure 10. Construction of Brickworks Shopping Centre.

3.2. Project Sustainability and Recognition

At the 2020 Master Builders Victoria Excellence in Construction Awards, the project won the Best Sustainable Project Award. More importantly, the project achieved full accreditation under the Living Building Challenge®, the world's most rigorous measure of the BE administered by The International Living Future Institute (ILFI). In the 'Materials' category of this accreditation, the project must feature at least one salvaged material per 500 m^2 of gross building area.

The mega-sustainable retail complex is branded as the "world's most sustainable shopping centre" and falls under the GBCA's Green Star (6-star) 'global leadership' category, which features 'recycle and store excess water on-site' and 'use natural resources to power the building.' The project has earned several other awards to date (Figure 11). The project was awarded Australia's Best Sustainable Development – New Buildings and Best Shopping Centre Development for 2021; the Rider Levett Bucknall Australian Development of the Year; the Landcom Award for Best Sustainable Development - New Buildings; and the Hames Sharley Award for Best Shopping Centre Development.

Figure 11. Project sustainability accreditations and awards achieved to date.

During the first phase of the construction project, the soil played a significant role. The whole location was once used as a quarry, with the deepest pit measuring around 35 m in depth. This meant that deep trenches already existed on the site and that they would need to be filled in, while at the same time, a new area would need to be constructed for the subterranean parking of the new retail centre. The soil on the property was moved to create a place for the parking lot while also contributing to the infilling of the quarry's deepest hole. The remnants of the old quarry were redeveloped into a parkland area with ponds and open space for future residents and tourists to enjoy. This replenishes the ground and invites nature back in, a core principle of the Living Building Challenge®. Extra soil that could not be accommodated on site was taken to other nearby construction sites to be used as infill.

3.3. Application of PwRC in the Case Project

While the target set by the client was to use more than 80 different materials and 30 application areas, most of the PwRC in this project were utilised in architectural features and civil works. As part of the Living Building Challenge competition, the project team had to use one PwRC per 500 m^2 in the project. The following are some examples of the PwRC application involved:

- The use of crushed concrete in the subbase of bitumen
- The use of materials from slab form working as hanging timber and timber cladding in the ceiling
- The use of second-hand brick purposed into tiles and concrete in floors
- The use of crushed brick leftover as a finish on facades.

3.4. Descriptive Findings: Participants' Profile

In this case study, interviews with representatives of the four main stakeholder groups were conducted between May and July 2022. The full description of the research approach employed is provided in Chapter 2. As tabulated in Table 5, the interviewees were quite experienced in their respective fields of expertise. The client rep (C_1P_1) had 20 years' experience in project development and notably environmentally driven buildings with a proven track record of successful interaction with many C&D contractors over these years. C_1P_1's expertise in specifying different construction materials including SMs/PwRC makes this participant aware of the sustainable management of recycled C&D waste products.

The head-contractor (C_1P_2) rep's experience in the organisation included six years of working in construction processes including direct engagement in product specification and on-site C&D waste management. The principal architect of the project (C_1P_3) had 11 years of experience with a wide domain of architectural-based activities ranging from private dwellings to large public facilities and commercial buildings. C_1P_3 was directly involved in projects where a mix of raw, second-hand and PwRC were used. The last interviewee, who represented the project's PwRC supplier (C_1P_4), has worked in their organisation for seven years. The participant's role includes selling recycled timbers that were repurposed from demolition timbers imported from demolition sites managed by their organisation or other demolition contractors.

Table 5. Summary of participants' profiles in Case Study 1

Stakeholder type	Experience/expertise	Participant code
Client	20 years' experience in construction project development	C_1P_1
Head-contractor	About 6 years' working experience in the organisation	C_1P_2
Architect	11 years' experience in architectural management	C_1P_3
Supplier	7 years' working experience in the organisation as the sale manager	C_1P_4

3.5. Organisations Involvement in the Project

Table 6. Main involvement areas of study organisations in Case Study 1

Organisation	Involvement/activities
Client	Preparing the tender documents, supplying the project funding, and working with the head-contractor to ensure that Living Building Challenge requirements are met.
Head-contractor	Procuring waste resources with the help of subcontractors and ensuring that they have the quality required for application in the project
Design	Providing design consultancy to the client regarding the use of PwRC prior to and during the construction phase
Supplier	Extracting timber waste from demolition sites and supplying salvaged timber to the project as per requirements

Table 6 summarises the main involvement areas of the study organisation in Case Study 1. The main activities of the project's client were to prepare the tender documents, supply the project funding and select the head-contractor. In terms of using PwRC, the main responsibility of the head-contractor was to review all waste resources that were installed and to inspect them for the ingredients to ensure only approved products were used. The design organisation was involved in the project at the planning and feasibility stage. The organisation provided design consultancy with respect to the utilisation of PwRC. The supplier organisation provided salvaged timber extracted from various demolition sites. This organisation processed the timber waste according to the client's needs and material specifications.

3.6. Stakeholders' Interaction Modelling in Case Study 1

Figure 12 illustrates a model that shows the contribution of each stakeholder and their interaction in using PwRC in Case Study 1. As can be seen in the figure, many interactions took place between the stakeholders to successfully apply PwRC and to meet the Living Building Challenge requirements.

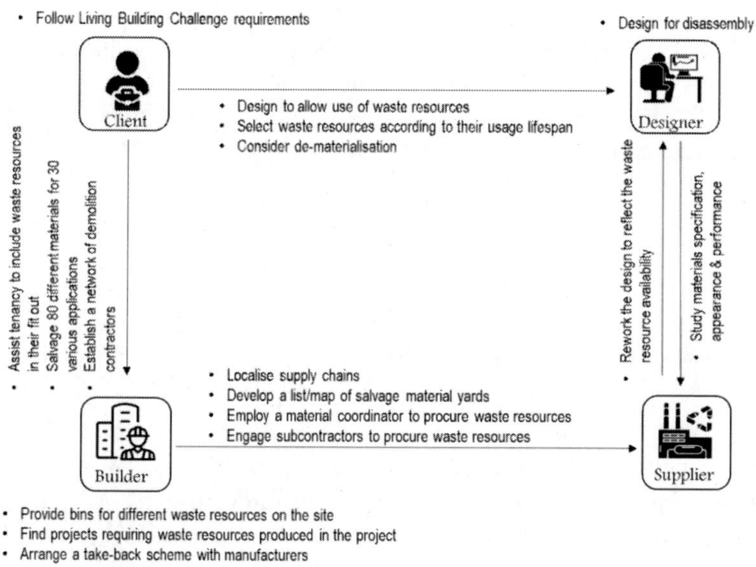

Figure 12. Stakeholders' interaction model in Case Study 1.

3.7. Major Motivations for Using PwRC

With regards to the main motives for using PwRC, and participation in this project in the case of the supplying organisation, the study stakeholders pointed to different factors. These factors are visualised in Figure 13 as per each participant interviewed. The following sections further describe these factors.

Figure 13. Main motivations for using PwRC in Case Study 1

C_1P_1 mentioned three reasons as to why PwRC were used in this project. First, the use of SMs brings about more health benefits compared to raw materials. The participant believed that the potentially toxic materials in the case of second-hand materials are gradually diminished. Second, the spiritual character of SMs and the associated attractiveness justify their usage over other brand-new materials:

> 'A lot of the salvaged materials have kind of a spiritual quality to them. So they have character and soul and a story. For example, we salvaged from local buildings or buildings around Australia that were being demolished or being refurbished. We could take, you know, parts of those buildings and then reuse them.' [C_1P_1]

Third, in some cases, the procurement of SMs/PwRC costs less than virgin materials. According to this participant, some of these materials were procured from the vicinity of the project and localised supply chains were activated. C_1P_2 indicated that the reason for SMs/PwRC was the combination of four factors: cost, social image, contractual obligation, and stakeholders' input and suggestions. The participant mentioned that the client's direction was towards establishing environmentally friendly product projects, which had seemed to be a difficult objective to achieve in the beginning but became easier as the project progressed:

> 'That was a contract or obligation...and I suppose at the start, it was hard...and just felt like a contractual obligation, like, this is 30 items is a really hard thing to do. But as you get the ball rolling, I think you get more creative, you start brainstorming things, you start looking at different ways...you sort of get that momentum run and you're happy to

put it in as many things as your client and architect are ready to take off. I think at the beginning of the job, it started off as a contractor obligation. That's probably the predominant reason but, yeah, it evolves as you go on that journey of the project.' [C_1P_2]

The social image that the organisation would build by being involved in this project and using PwRC was a major motive according to C_1P_2. The participant also indicated that suggestions from various parties provided further motives to use SMs/PwRC in different applications:

'I suppose from social image and as a direction that the company sort of wants to go and we want to be exposed to more environmentally friendly product projects. So, as a whole, that project was something that we were really interested in from a business level.' [C_1P_2]

'Having a brainstorming session with client consultants helped as well. That was sort of the main considerations, we also had input from subcontractors...there's lots of it [architectural features around the travelator] that looks like just hanging timber and timber cladding. But that's actually all formwork from the slabs that we poured...we actually got it, stripped it back and reuse it pretty quickly in the building. So that was a subcontractor suggestion that I put forward.' [C_1P_2]

The design team rep (C_1P_3) indicated that the use of PwRC was proposed to assist the client's attempt to fulfil the requirements of the Petal Certification (Living Building Challenge). According to the calculations done by C_1P_3's organisation, the use of SMs/PwRC (e.g., recycled hardwood timber) resulted in a positive project's carbon footprint due to sourcing them from shorter transport distances.

'So obviously, we liked the idea from the start, but I think the certification process assisted with quantifying what needs to be achieved. In this case, we had a carbon footprint calculation and transport was part of that conversation. So, transporting fresh hardwood timber had a much larger impact on the mileage and that's why this have shouldn't in the end, it was a balancing act because all the figures had to balance out. And in this case, the hardwood was significant assistance to achieve that outcome.' [C_1P_3]

The research participant from the supplying organisation (C_1P_4) listed a few reasons for engaging in the project. These include prior familiarity with the builder, the opportunity to work with a reputable builder and the recognition they received from participation in this environmentally friendly project.

> 'That [Project] is pretty impressive...we were really excited to be on board with something like that. I mean, we hold ourselves, especially being part of the recycling part. In terms of our recycling and stuff like that our reputation, so it was just, there was a no-brainer to want to be part of it. And as I said, HACER were amazing. In every way, and even the builders that were a part of it as well…it was a good thing to be part of and the recognition that we did get from it was really good as well.' [C_1P_4]

3.8. Major Barriers to Using PwRC

The use of SMs/PwRC in the Brickworks Shopping Centre project came with some challenges, which are presented in Figure 14. The analysis of responses showed that the availability of SMs/PwRC was the most frequent reference issue as indicated by the builder, designer and supplier.

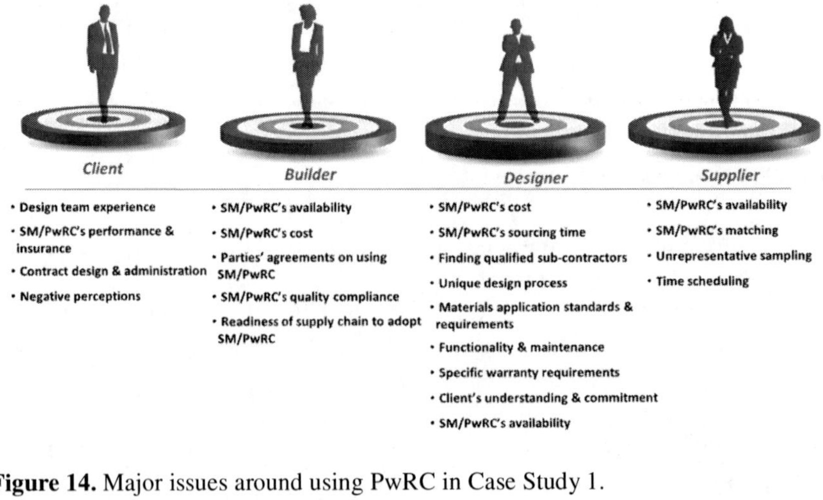

Figure 14. Major issues around using PwRC in Case Study 1.

C_1P_1 listed several challenges that included issues around the design team experience, the performance and insurance of the materials, the contract design and administration, and negative perceptions. The participant indicated that due to the unique feature of the project, designers need to do things that are not routine, which also hindered the design and administration of a contract that requires the effective collaboration of contractors to use these materials:

> 'For example, an architect's might design a room and then the contractor tries to find recycled materials to use to actually build that room. But…we found it was easier to find what materials are out there in the market first, and then have a palette of materials and then do the design, you know. And I think that's a challenge for our industry, because we don't generally work in that way, we tend to design something, then try to procure materials to make them fit into what's already been designed, rather than understanding what our materials opportunities already are.' [C_1P_1]

Furthermore, these materials (e.g., salvaged hydraulic tapware, showerheads and taps) might not perform as well as their brand-new alternatives so they might not pass the requirements dictated by relevant building codes and standards. For this reason, their insurance and guarantee requirements differ from those of brand-new materials. Lastly, the negative perception towards SM/PwRC embedded in wider society leads people to associate these materials with lower quality:

> 'It's usually the perception of the end users, but also sometimes the tenants themselves, they kind of some of them are the end users, right? So it depends on the tenancy type, but they will have their own, like shop fitter or their own contractors to do the layout work and their own contractors have their own prejudices as well, you know, saying, Oh, if I use this recycled product, do it won't last as long or it is not as robust or whatever. Maybe that's not that's not the case. It's just usually a perception.' [C_1P_1]

The builder representative (C_1P_2) enumerated five challenges of using PwRC in the case project. These include the availability of SM/PwRC, agreements of all parties on using the proposed PwRC, the cost of the materials, quality compliance with the relevant policies and specifications, and readiness of the supply chain to adopt these materials. The participant reported that it was difficult to procure a wider range of SM/PwRC in the high quantity required for the project. Also, the task of getting the agreement of everyone involved in the project (i.e., client, designer and sub-contractors) to use the PwRC was seen as a challenge by C_1P_2, echoing the fact that the supply chains are not well developed to adopt these materials:

> 'So, we would come up with something that we think is great, and the architect would look at it and think you're out of your mind. That's disgusting…it's just making sure that your subcontractors are okay with it. They're the ones that have to install it and sign it off at the end of the day. So, they have to be comfortable with the product that they're putting forward that they can give you the warranty that we need and those sorts of things that those who had contract obligations are still met using a product that they're unfamiliar with.' [C_1P_2]

The cost of SM/PwRC was reported to be another issue; according to C_1P_2, procuring uncustomary materials of a certain quality with no defined cost limit as reflected in the specifications made it unaffordable to their organisation. In some cases, this limitation resulted in procuring of alternative materials in this project. Furthermore, the participant commented on the difficulty of achieving quality compliance with the relevant regulations. This is the direct result of the lack of traceability of PwRC procured for the project. According to this participant, sometimes the procured SMs that do not have certification from a recycling facility or manufacturer did not have the intended performance quality or simply did not fit for the purpose:

> 'If you're buying something that's being reused, and it's from the second-hand yard, or that sort of thing, that was definitely tricky. Even after the carpet was one, we used some carpet there, that was reused. But we haven't had trouble with those peeling quarters and not being able to fix them down. Because it's not really flat, it's been pulled up from a previous job. So, laying that down again. So same sort of thing, just the quirks I guess you get with a salvaged product, it's not you and you have to sort of manage that.' [C_1P_2]

C_1P_3 reported that the design team faced several challenges when preparing design work to use PwRC in this project. The participant pointed to cost, time, finding qualified sub-contractors, unique design process, the material application standards and requirements, functionality and maintenance, specific warranty requirements, the client's understanding and commitment and availability of materials. According to C_1P_3, the application of PwRC is typically more expensive and needs extra resources for double handling. The design team also needed to spend more time searching and sourcing various PwRC alternatives. Furthermore, not all sub-contractors were able to handle uncustomary PwRC and finding the right subcontractor was a challenge in this project. The design team had issues with the specification of some PwRC proposed as they were unable to pass some ratings imposed by building codes and internal guidelines or simply did not have the desirable functionality and maintenance requirements:

> 'So, for instance, the recycled timber ceilings, they have to follow a certain fire group rating and in order to achieve that, you have to get a certain thickness, and you have to sign it off... So that goes back to your fire group ratings, because if you have got a good strategy in place, you might be able to have a slightly reduced assessment of let's say, the ceiling treatment. So there's sort of mechanisms that you can use to get it through the code that in the end is obviously dictating the compliance…it would have been probably not necessarily regulatory, but also just in terms of functionality and maintenance, that while something might look good, and might have a narrative around it being recycled, and a reduced carbon footprint, the realities of it being in an area, which is exposed to a lot of wear and tear, and a robustness requirement for its functionality would have been a serious consideration as well. And that's, you know, why we often go with new products, because they're more and more robust, because they're tested really thoroughly.' [C_1P_3]

The other challenge in C_1P_3's opinion was to clarify the client's commitment and understanding of the consequences of using PwRC, such as the lack of warranty in the case of brand-new materials:

> 'And your clients got to understand then, because it's a client who's then holding the accreditation for the Living Building Challenge, which is that which was one of the pushes to use a lot of the recycled material, reduce their carbon footprint. So yeah, it was a commitment from them as well to understand warranties are not the same as some of the other commercially available products. I think that's a really important point for Cinder because, that not every client is willing to do that and I think if you identify risk, I think warranty with recycled products is probably very, very, very hard.' [C_1P_3]

The participant indicated that the procurement of these materials with similar qualities in sizable quantities was a challenge. These materials also warranted a unique design process, meaning that the design team had to come up with a preliminary design, which later had to be checked with suppliers of PwRC:

> 'I know, we're talking about a single building here. But a lot of these things are much easier at a much smaller scale and you don't need them in the same amount of quantities. So sourcing things was not just finding the product is finding enough of the product, because you're actually using it for a lot of larger space to actually…So it's quite difficult to reuse recycled objects much easier at a domestic scale than it is at a commercial scale…Because they were looking for very specific things like pre-Mac doors that were completely recycled windows And so there had to be a, it had to be an initial design, then a search for those particular components that were recycled. And then another redesign based on the parameters of those found objects. So that change, there had to be a kind of unfinished design concept to get through that process.' [C_1P_3]

The C_1P_4 also highlighted the challenges of using SM/PwRC in the case project. These were the difficulty of supplying enough materials, matching the materials, unrepresentative sampling and time schedule. The participant mentioned that, due to the large scale of the supply, sometimes it was difficult to get the quantity of materials required and, in some cases, due to the low quality of a fraction of the feedstock, matching the materials was difficult. Furthermore, a few samplings did not give assurance of the final look of the products when used in the project. They described:

> 'You might have the feedstock there. But when you're machining it, there may be hollow bits in the timber. So which means it collapses, and then you can't get the material that you think you're going to get out of it. So I guess the machining side was a challenge…So I think the colour tones and all that kind of stuff was the only thing that was the challenge, but because of the large scale of the material, you can't make it all look the same' [C_1P_4].

The other challenge was to achieve a time schedule plan. The supply of the PwRC (timber in this case) was done in different milestones which makes the matching practice more difficult. The participant explained:

'The scheduling, just making sure we got everything out on time, because it wasn't just, it didn't all just go out at once. Obviously, there because of the quantity it was done in Stages.' [C_1P_4]

3.9. Major Strategies to Overcome the Reported Barriers

Several strategies were proposed by the participants that would have improved the use of PwRC in this project (Figure 15). C_1P_1 indicated that some strategies such as clear requirements for using SM/PwRC in contracts, normalising their usage across the supply chain, legislating SM/PwRC usage and imposing highly rated landfill taxes on waste resources would be helpful.

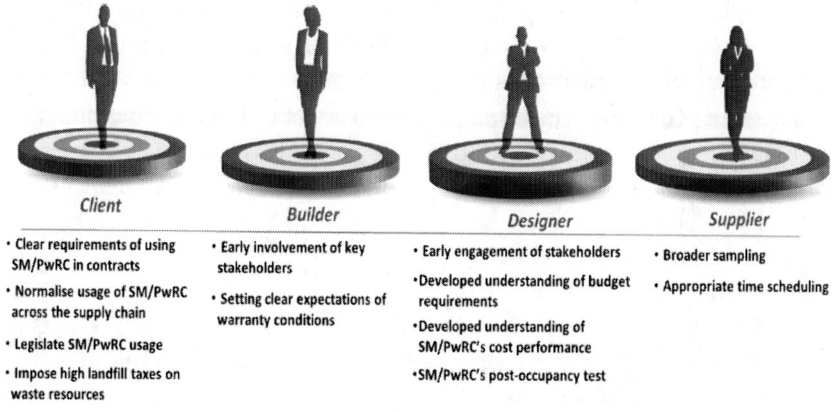

Figure 15. Major strategies to remove barriers to using PwRC in Case Study 1.

According to this participant, when the contract was clear about the requirements for using PwRC, the parties involved were prepared to work with them. Preparation for this involves creating the capacity to consider these materials in different ways. C_1P_1 also indicated that when applying CE, the entire construction lifecycle (i.e., design, construction and operation) should be considered. The participant also reported that although it might not be straightforward, having the application of these materials legislated would remove some of the barriers. Lastly, C_1P_1 mentioned that having a high rate of landfill tax on C&D waste resources will enhance the uptake of these materials in construction projects. They described:

'The second element is like really working out how to drive that circular economy in a way that isn't said, a lot of people are happy to give to the circular economy by like, providing old things, but they won't buy them back themselves. Right. So, having hat whole lifecycle embedded into design, construction, but also operation like why? Why are we not using products, even in the operation of buildings, not just when they're being constructed?.' [C_1P_1]

'If I'm really, really honest review a really high landfill tax. I think we've changed the way people behave. At the moment, it's almost free to pollute.' [C_1P_1]

C_1P_2 reported that overcoming the challenges of using SMs/PwRC can be planned through the early involvement of key stakeholders and setting clear expectations of available warranties for these products. Early involvement of the stakeholders through establishing a framework would help everyone to understand what the final outcome could look like. Furthermore, this participant argued that there must be clear expectations of warranty requirements of these products that vary from brand new ones and can create a positive environment among stakeholders in favour of using them. Clarification would help adjust stakeholders' expectations not to overestimate the performance of these products:

'I think a lot of these is having really early and open involvement with everyone on the project. So, it's setting that expectation, I think with your client about what they want, again, having an architect on board early with even just some ideas of what they would like to have either recycled or salvaged. So you're not starting from scratch and you can sort of set that expectation with what they would like what an absolute NO.' [C_1P_2]

Participant C_1P_2 highlighted the role of reasonable timing and budget, client's readiness to accept the realities, early engagement of key stakeholders, and awareness of SMs/PwRC performance as the main strategies in removing the barriers reported by them. According to this participant, having a client with a developed understanding of the specific requirements of working with these products, such as allocating reasonable timing and cost uplift, would reduce the impact of the challenges reported before. Notably, due to the uniqueness of projects using SMs/PwRC, time was emphasised to be a critical factor in the design process:

'You need a client on board who's willing to pay for it. It's not cheap. When it does take its toll on people in terms of time as well…I think one thing there's often a misconception about a lot of architecture and that is just that we do whatever we want. And that if we say we want recycled materials, they just get into a project and it's not, it's a conversation and a journey with the client.' [C_1P_3]

'I would say more time and more money, but time is money. So at the end of the day, like that's fundamentally like with any of these projects, any sustainable kind of strategies, whether it's recycled materials, [or] any of it…the time one is quite critical as well, to really, obviously, that's related to money. But it's allowing for those processes to occur. And often, these projects are connected to an opening of a major tenant and that deadline can absolutely shut down your design process and if you're going to have these large tenants, you would say straight up, before it even starts, you need to make sure that you allow enough time to go through this process. Because then it might not all be cost. So some materials are more expensive, but it's the time it takes to certify, or the time it takes to calculate it all the time it takes to sources.' [C_1P_3]

C_1P_3 mentioned that early engagement of stakeholders is a must in this kind of project. Notably, it allows them to develop an understanding of budget requirements associated with using these products:

'Like early engagement of client and understanding and the supply chain, obviously, for our supply chain and their budget consideration, because the thing is if a recycled material is 1.5 times the price of something that is not recycled, any client who is not committed to that strategy will take the cheaper option. Definitely a more reliable option in some cases.' [C_1P_3]

Another point made by this participant was about developing knowledge regarding the cost performance of using these materials. C_1P_3 mentioned that the cost uplift when using these materials needs to be considered in the long run, which could have a significant impact on the decision made by clients about to whether use SMs/PwRC:

'[What] I do want to add is the cost uplift that sometimes occurs through these decisions is not necessarily a cost uplift, if you assess a building over a 30 or 50-year time period. So that's, I think, to be made quite clear.

Because when we talk about resilient and regenerative buildings, you have a performance and a much smaller footprint, and a much lesser, let's say, energy bill. If you measure that with just your construction costs, and let's say five years, yes, it's a more expensive building. By the time you get to 20 years, you have a cheaper building. So that's something worth maintenance cost if you want to compare the raw materials.' [C_1P_3]

The participant added that having post-occupancy test for each material would facilitate precise calculation of cost implications of using PwRC through comparing their benefits versus maintenance cost:

'...we were starting to see what the performance differences are like in terms of electricity use, recycling water and all of those things. But in terms of the maintenance cost, you would think that hardwood and timber would last quite well. For the brick, you know, that's been used for a long time. And should last as well. It depends on the material, it needs a post-occupancy assessment set, you also need time, and this is a very new project. So your project exactly that's, something that not many people, not many businesses know about it. So that's, really quite new initiative. So we still trying and testing things. And still a long way to go to understand the issues ahead of us.' [C_1P_3]

The participant representing the supplier, C_1P_4, noted that taking broader sampling and appropriate time scheduling would be two strategies to overcome the main challenges of using PwRC in the Brickworks Shopping Centre Project. More frequent and precise sampling from PwRC sourced from different sources would provide a realistic view of the quality of the final work applied at a large scale. Having clear, mutually agreed upon, and reasonable time scheduling was also noted to be an important factor when handling PwRC:

'I think with sampling and stuff like that, which I think we did do but is maybe give them a broader example of what it can be. Because at the end of the day, it is a natural product, especially with timber. Also noting that there can be movement with timber, it's a live product, it does move so it moves to its surroundings.' [C_1P_4]

'It can be a bit stressful when you've got to make sure you've got X amount going out because they've got people on site that and they need to put this wall up or they've got that out. Planning and you know what I mean? You got to get the staircase finished or yes, just, I think the scheduling side of it, making sure you organise from everyone's end.' [C_1P_4]

Chapter 4

Mordialloc Freeway Project (Case Study 2)

4.1. Project Background

The Mordialloc Freeway Project (MFP) in Victoria comprises a 9 km freeway link between Dingley Bypass and the Mornington Peninsula Freeway, including several grade-separated interchanges, six new road bridges over wetlands and Mordialloc Creek, and over 8 km of parallel shared user path along the alignment. The MFP was built between 2019 and 2021 using the delivery method design and construct (D&C). This Major Road Projects Victoria (MRPV) freeway was constructed by McConnell Dowell and Decmil Joint Venture (MCDDJV). The partnership intended to deliver a world-first freeway (Figure 16), one that would incorporate technological advancements in sustainability, delivering not only a long-awaited stretch of road, but also contributing to the state's waste reduction goals.

4.2. Project Sustainability and Recognition

The project has delivered several social, economic and environmental benefits. In terms of the traffic outcome, according to MPRV prediction modelling, the project will save 7 and 10 minutes in peak morning and evening traffic, respectively. It will also reduce 60–70% of the traffic load on Wells Rd and Boundary Rd by 2031. To drive the social outcome, the project team also implemented many social procurement initiatives and achieved the following:

- 92,702 total hours for local content and employment outcomes;
- 49,242 total hours for Aboriginal employment; and
- 71,068 total hours disadvantaged Victorian commitment.

To date, the project has been recognised for its environmental sustainability, contractor performance and design. As can be seen in Figure 17, the project has received awards between 2021 and 2022.

Source: Victoria Big Build 2022 [56].

Figure 16. Construction of Mordialloc Freeway.

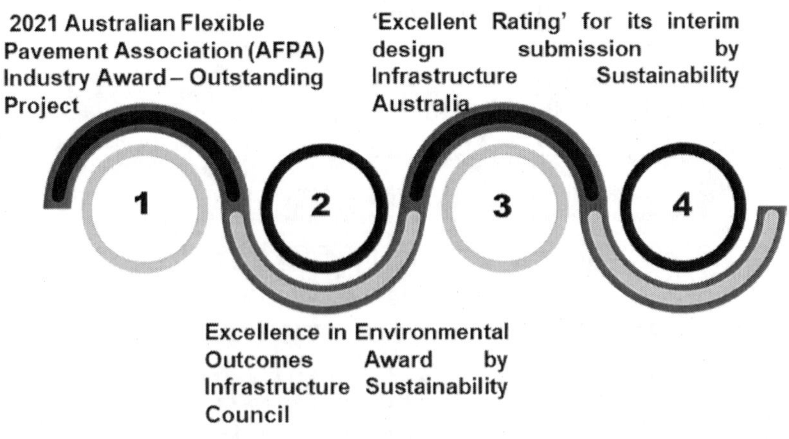

Figure 17. Project sustainability accreditations and awards achieved to date.

4.3. Application of PwRC in the Case Project

The project had significant environmental impacts. Since the project's completion, the MFP has been known as 'Australia's greenest freeway.' This claim is mainly supported by a collective of implemented sustainability initiatives in response to Victoria's waste crisis as highlighted in the report *Recycling Victoria: A New Economy* [66]. The project saved over 300,000 t of materials from landfill or 3 ha of land space that may be required for stockpiling. The use of these products compared to comparable products on the market has saved over 4,000 t of greenhouse gas emissions.

The PwRC quantity, type and application areas in this project include the following:

- 75% recycled plastic rotationally moulded panel (RMP) noise walls (10 km of noise walls using 600 tonnes of plastic waste equalling 32,000 m^2)
- 100% recycled subbase pavement materials and 44% PwRC in asphalt pavements (over 270,000 t of pavement material incorporating the maximum allowable recycled content)
- 10% crushed glass and 30% reclaimed asphalt pavement in pavements
- 100% recycled polypropylene plastic (PP) concrete reinforcing mesh (30 t of plastic waste)
- 100% recycled high-density polyethene (HDPE) stormwater drainage pipe (4.6 km installed totalling 75 t of plastic waste)

The project applied 407 VicRoads Standard Specifications [61] to ensure the quality of PwRC to be used in this project met the minimum acceptable range. This specifications document was updated in collaboration between the industry and government agencies to allow for more recycled asphalt products to be used in all pavements going forward.

4.4. Descriptive Findings: Participants' Profile

Four key stakeholders of this project were interviewed between June and August 2022 (Table 7). The full description of the research approach employed is provided in Chapter 2. The interviewees were experienced individuals with a significant track record of involvement in building public infrastructure in Victoria. The client representative (C_2P_1) was extensively engaged in the Mordialloc Freeway Project in two capacities: project director at the pre-construction and procurement phases, and senior project manager responsible for the delivery of the southern part of the project. The participant representing the head-contractor (C_2P_2) has worked in the building and construction sector for the last 20 years, and experienced different roles from site engineer and project engineer to project manager in various road infrastructure and mechanical projects. As the project manager, C_2P_2 led a group of engineers working on different parts of the project. This participant's involvement in C&D waste management was limited to engaging waste operators to take the on-site waste to landfills.

The third research participant (C_2P_3), who represented the design team, had more than 15 years of experience as a design consultant, primarily on road infrastructure projects such as the removal of PTV level crossings and the construction of the West Gate Tunnel. This particpant's primary experience with PwRC was related to the case study project. In this project, C_2P_3 managed the main freeway alignment packages, the freeway crossings with major local roads and the design coordination in the role of civil design lead. The final participant (C_2P_4) worked as the organisation's communication manager and participated in the project development process at an early stage. While the participant was not directly involved in managing the project's PwRC supply, they had a good understanding of the organisation's supplying activities due to their participation in some communications with other stakeholders.

Table 7. Summary of participants' profiles in Case Study 2

Stakeholder type	Experience/expertise	Participant code
Client	Senior project manager with extensive experience in project managing public infrastructure projects	C_2P_1
Head-contractor	20 years' experience in the building and construction sector as a site engineer, project engineer and manager	C_2P_2
Designer	15 years' experience working as a design consultant	C_2P_3
Supplier	Highly experienced corporate communicator with a background in government, corporate, industry and community organisations, with a four-year employment history in the organisation	C_2P_4

4.5. Organisations' Involvement in the Project

The client worked with the PwRC suppliers to increase their application in the project leading to the revision of materials specifications. The client had meetings with the project's potential contractors and suppliers to review the project scope and requirements. Furthermore, the client project team established the sustainability committee to monitor and report the use of PwRC in this project. The head-contractor that represented a joint venture was responsible for conducting the construction activities according to the design. This organisation also procured PwRC from various suppliers.

The design team was responsible for the design work mainly in the civil roads' drainage, structural design, geometric highway, traffic management and alignment of the other roads. The supplier had a key role in suggesting reforms to the existing PwRC specifications in this project. The main materials they supplied included reclaimed asphalt pavement, crushed glass, crushed rock and concrete. A summary of the stakeholders' main involvement areas in the project is presented in Table 8.

Table 8. Main involvement areas of case study organisations in Case Study 2

Organisation	Involvement/activities
Client	Working with stakeholders to develop new materials specifications and establishing a sustainability committee to monitor sustainability outcome
Head-contractor	Conducting the construction work according to the design
Design	Providing design work for the civil roads' drainage, structural design, geometric highway, traffic management and alignment of the other roads
Supplier	Working with the client to revise PwRC specifications, supplying reclaimed asphalt pavement, crushed glass, crushed rock and concrete

4.6. Stakeholders' Interaction Modelling in Case Study 2

Figure 18 shows the interaction model between the four study stakeholders in Case Study 2. The modelling was done based on the responses captured during the interviews.

4.7. Major Motivations for Using PwRC

The four stakeholders interviewed in this case study provided 20 reasons why they have been engaged in this project and why they were motivated to use PwRC. The most frequent reason was to contribute to the environmental sustainability of the construction activities, which was indicated by the client, designer and supplier of PwRC. Figure 19 illustrates these motivations for each interviewee.

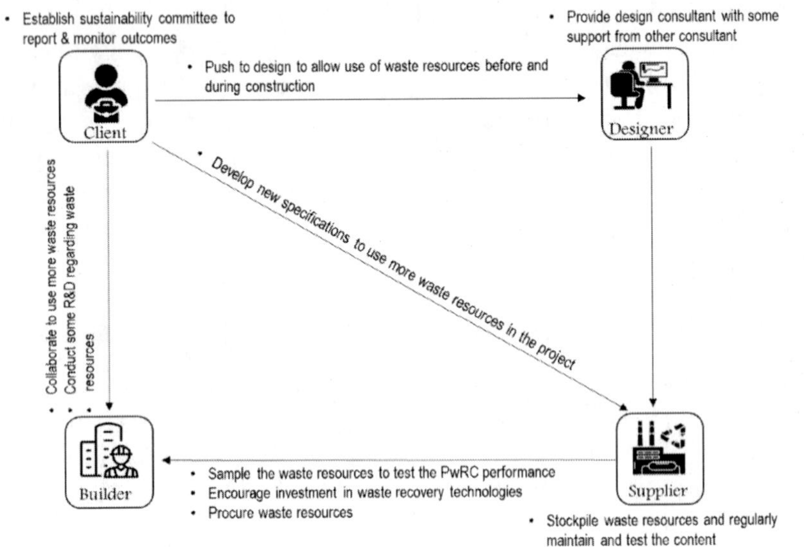

Figure 18. Stakeholders' interaction model in Case Study 2.

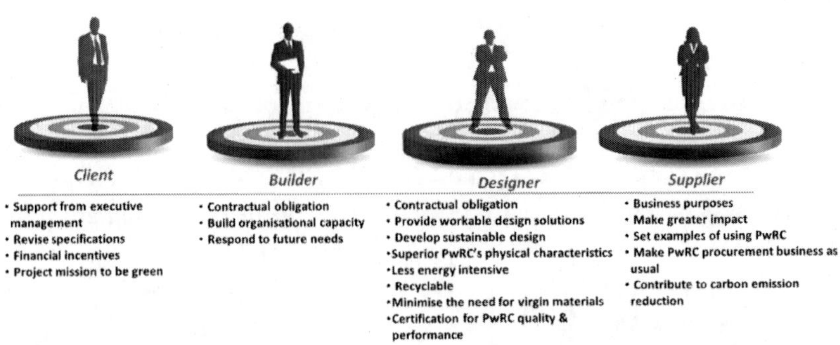

Figure 19. Main motivations for using PwRC in Case Study 2.

The client representative reported that support from the organisation's executive management, revising the materials specifications, financial incentives and the project's aspiration to be green were strong motivations to use PwRC in this project. C_2P_1 highlighted that their executive management team was a strong supporter of using these materials as it would support recyclers, reduce the quantity of waste landfilling and contribute to the state's economy:

> '...to be fair, if I had been just the project team, we might not have done that, but we were supported and encouraged to do that from our executive. And that was done on the basis of supporting recycled materials, but also supporting the Victorian economy.' [C_2P_1]

Furthermore, according to this participant, the DoT specifications were revised in collaboration with the supplier to allow more usage of PwRC in this project, notably asphalt and plastic noise walls. The client organisation also provided financial incentives for the head-contractor to achieve the targets set for using PwRC in this project, which was based on a 70% ISC rating:

> 'Ultimately, we worked with the DoT and industry...to develop a new specification, and we used it on the Mordialloc freeway project. And that allowed for more recycled asphalt products to be used in all pavements going forward...that opened up a lot more recycled products into the Victorian market. And that was the big deal with the Mordialloc project.' [C_2P_1]

> 'What we also did is we also provided a financial incentive in that contract to achieve a 70-rated ISC outcome.' [C_2P_1]

> 'I think it was a project goal to build the greenest freeway that we could. So there was a motivation, from the organisation from the individuals who planned that to achieve that...there's benefits in relation to the Victorian economy to provide all the support producers that we're in Victoria. So the recycled first element of that, you know, to reduce the, you know, the, the amount of product going to landfill.' [C_2P_1]

C_2P_2 referred to three factors as the major enablers of or motives for using PwRC in this project: contract obligation, building organisational capacity to work with PwRC, and preparing to respond to current and future client needs. In the contract awarded to the head-contractor organisation, the use of PwRC was required to meet the ISC sustainability rating requirements. Furthermore, according to this participant, with the emerging trend in the state's public projects to use more PwRC, the participant's organisation willingly adopted these materials to build the organisational capacity to better address the market needs and secure the kinds of projects in which the use of PwRC is necessary:

> 'Because it's good for business, you need to differentiate yourself from the others and then you need to respond to your clients to know what they want, what they need out of projects. This is when we sell ourselves for the next project. What we're used to as our selling point is we say well, look at all this great stuff, we are a company that does this type of thing as part of our business models.' [C_2P_2]

C_2P_3, the representative of the design team highlighted seven factors as the main motives behind using the PwRC. These include responding to the client's request, providing design solutions that satisfy all stakeholders involved in the project, developing sustainable design, the superior physical characteristics of PwRC, being less energy intensive, being recyclable, minimising the need for virgin materials and certification for their quality and performance. The participant explained:

> '…we definitely sort of wanted to provide a good sort of sustainable design. That was one factor of it. It was just about really keeping the client happy, as well as the contractor and coming up with a solution that worked for everyone.' [C_2P_3]

This participant emphasised the environmental benefits of using PwRC in their project. The benefits realised included reducing the need for using virgin materials, consumption of materials that consumed less energy to be produced and a higher level of recyclability at the end of their service life, which further contributes to a CE in the building and construction sector. The participant expanded on this, stating:

> 'For the noise wall panels, for example, I know the benefits of using the recycling plastics is that they're lightweight, they've got the certification behind them in terms of testing for UV resistance and chemical resistance, they're less sort of energy intensive to make and manufacture. And then you can also easily replace sections of the wall and also recycle them at the end of their useful life as well. So you can do, all those things also with other materials such as concrete and steel. But you're already

starting with a recycled product to begin with, rather than creating something, and then going on to recycle it later down the line.' [C_2P_3]

C_2P_4 also highlighted three main reasons for being involved in this project. These reasons were business purposes, and to make a greater impact, set examples for other smaller council and private construction projects, make procurement of PwRC business-as-usual and contribute to lowering the sector's carbon footprint. The participant's organisation had been supplying PwRC with construction projects for many years, particularly in Victoria and Queensland. Hence, their engagement in this project was primarily based on business grounds:

'...we probably the largest recycler, by weight, and volume in the country, we recover materials to manufacture heavy materials for roads and infrastructure. So this is our bread and butter. And we would always seek to supply major projects with these materials, because that's where we can make the biggest impact.' [C_2P_4]

Furthermore, according to C_2P_4, by engaging in this project, they were able to make the greatest impact due to the scale of supply, which further resulted in building confidence in the market to use these materials, particularly in smaller scaled construction projects. This eventually should result in using these materials as business as usual. This project allowed them to set an example for the safe application of PwRC in the Australian context. Lastly, the use of PwRC contributes to reduced carbon emissions resulting from construction activities. They explained:

'...we want to make sure that major projects especially are getting, creating, or paving the way for councils and smaller construction, and setting an example. We know that through the use of our materials, these projects can reduce their carbon footprint by up to about 65%. So having these materials used in major projects is a really great way to communicate to the rest of the market, that these are really viable materials to use.' [C_2P_4]

'...but we want to make our motivation for this...to make this part of business as usual for the entire sector. Purchasing and utilising recycled materials and infrastructure should just be the norm.' [C_2P_4]

4.8. Major Barriers to Using PwRC

In response to the question regarding the primary challenges associated with the use of PwRC, the respondents identified several factors that inhibit the widespread adoption of these products. As illustrated in Figure 20, unsupportive specifications and availability of PwRC are the major barriers reported. These two factors were indicated by all participants as the main two major barriers.

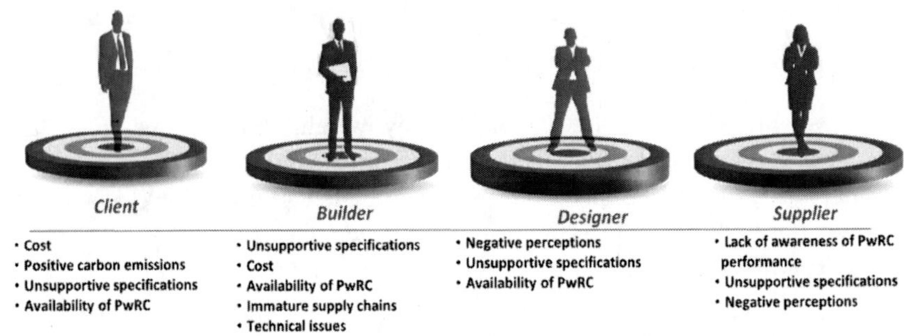

Figure 20. Major issues around using PwRC in Case Study 2.

Cost and positive carbon emission, as well as specifications and availability, were cited by the client as the most significant challenges in this project. C_2P_1 indicated that the cost of acquiring PwRC was greater than that in the case of new materials due to the small scale of production:

> '...the cost was one [challenge] and we did take the choice to pay a premium to move from a very efficient product in the noise walls, to the recycled plastic product.' [C_2P_1]

In addition, the participant noted that the use of PwRC would not necessarily result in a reduction in carbon emissions, which influenced the decision to use some recycled products. Lastly, according to C_2P_1, while the organisation was able to overcome some of the obstacles imposed by the Department of Victoria (DoT) material specifications, other restrictions following specifications prevented the use of additional PwRC:

> '...we weren't able to do that, ultimately, because the emissions for the trucks to pick up the water and take them to the site was greater than the benefit of actually using the recycled water [PwRC]. So yeah, there's a

balance in sometimes you get environmental benefits in one area, but you're harming the environment in another area. So, there were trade-offs that sometimes did not work in favour of recycled materials.' [C_2P_1]

'...so we had the contractor who wanted to achieve it [target of using PwRC], because they had commercial benefits. And then we had suppliers who were reluctant. And they were reluctant, because the DoT had these specifications...So whilst we were able to overcome some of those specification challenges with the asphalt there are probably other areas that we didn't even try.' [C_2P_1]

C_2P_2 mentioned five factors in response to this question: traditional specifications, the premium cost of PwRC, their availability, immature supply chains and technical challenges. According to this participant, the problem with conventional specifications is that they do not recognise the use of new materials in the common elements of infrastructure projects, such as the use of recycled plastic instead of concrete in pipes. It was also noted that the high cost of PwRC prevented the use of these materials in this project. The participant further connected this obstacle to the immaturity of the waste recycling industry and its inability to produce products at reasonable prices and the required scales:

'First and foremost is traditional specifications don't allow it...the pipe for instance. Well, the specification said it had to be concrete pipe. So if I was just to follow the specification that I can't do it all the recycled asphalt. Well, there was no testing regime there was not an accepted product. So the challenge is to get the specifications to be invited innovation specifications traditionally. They prevent innovation.' [C_2P_2]

'Some of these products come at a premium cost. And probably just because the industry is a bit immature, like they don't have the volume and the scale on some of these products yet, so I think we need to get more volume and scale and to get to drive the price down...there was only one supplier up in Mildura [that could produce sleepers with recycled plastic], and they could only produce like five a day or something which meant we could only do a very short section of the work, they just didn't have the capacity in their workshop to produce the amount that we needed. And they didn't have the money to invest in that production scale because not enough jobs were using it.' [C_2P_2]

The last challenge was related to the technicality aspect of the application of these materials. The participant noted that the characteristics of PwRC are different from new products, which created issues for their application in this project:

> '…for instance, the thickness of the panel causes any issue. So there's just that the maturity of the engineering that goes into them, so to know how to make the product work with all the rest of the systems that fit around it…And these noise walls were quite a lot thicker than traditional ones. So all of a sudden, you've got to know, from a design point of view, our designers left whatever was a two and a half metre space to fit all this in.' [C_2P_2]

The third participant, C_2P_3, indicated that the three major challenges were negative perceptions, specifications and the availability of PwRC. They described:

> '…there's a perception that the newer materials are a bit sort of untested. So getting the road authorities to approve the use of new materials can be a challenge. So obviously, with roadside infrastructure, it needs to be sort of performance tested, you don't want to instal something which is going to sort of fail or fall apart in the next 10 years, when it should be lasting for much longer…so there is some trepidation around that until it's been demonstrated that the product or the recycled material actually performs a similar function to traditional materials.' [C_2P_3]

> '…availability of recycled materials as well, it's probably easier in metropolitan Melbourne to get access to these alternative materials…but maybe some other parts of Australia, it's more difficult when you've only got a limited supply chain to deliver materials for projects.' [C_2P_3]

The last participant in this case study (C_2P_4), referred to the lack of awareness of PwRC performance, unsupportive material specifications and people with a sceptical mindset about PwRC as the main three challenges, stating:

> 'I think the challenging part for us is that there's a lot of talk about recycled materials being new or innovative or cutting edge…people wanting to do trials and things like that to research [application of] recycled materials [in] projects. The fact of the matter is that we've been

manufacturing these materials for decades, and they have been used for decades in projects across Victoria and Queensland. So we're actually past the point where this is new and newsworthy.' [C_2P_4]

'...the specifications are probably the key thing if the specifications allow for the use of recycled material and that takes away a huge barrier and makes it far easier to get acceptance for it in the market. So you know, it's open to the regulators and Vic Roads and major projects and people like that to really, you know, step up and make sure that there aren't these, you know, ridiculous invisible barriers in the way of people taking up the use o 'recycled material.' [C_2P_4]

'So if you have individual projects, engineers, or people working on the project that have old ideals around the use of materials that don't have experience with recycled materials, I might be a little bit resistant or reluctant to take that on.' [C_2P_4].

4.9. Major Strategies to Overcome the Reported Barriers

Different strategies were suggested by the interviewees to remove barriers to the use of PwRC in this project. As depicted in Figure 21, 12 strategies were captured in the interviews that can be clustered under three categories: financial, regulatory support and technical.

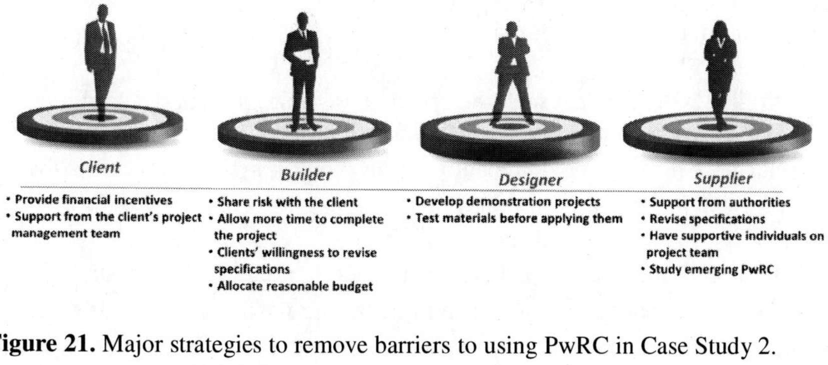

Figure 21. Major strategies to remove barriers to using PwRC in Case Study 2.

The first participant (C_2P_1) noted two major strategies: financial incentives and a supportive client project management team to monitor the sustainability progress of the project. They explained:

'...we had a sustainability committee for the project, which met monthly, we reported and monitored outcome outcomes in that timeframe as well...[we also] monitor the success in this in this area. So yeah, it was not treated as a as a side matter, it was treated as a significant element of the project that needed to be monitored and managed to be able to be achieved.' [C_2P_1]

The second participant named the main strategies to overcome the reported barriers as sharing the risk with the client, allowing more time than usual, the client's efforts in revising/updating materials specifications and the client's preparation to allocate a reasonable budget. They believed that taking away some risks from using these materials could result in the contractor's motivation to use more of them. Furthermore, the time allowed for exploring options and testing innovative ideas was seen as a way to solve the technical issues that contractors may face during construction. They stated:

'I think the clients need to take away the risk from the contractor on the performance of these innovations. So that would motivate us to use them a lot more. They need to give us the time to explore the options properly to look for the answer.' [C_2P_2]

According to C_2P_2, the client should be prepared to work with the contractor to revise the specifications that facilitate the utilisation of a wider range of PwRC with different characteristics. As the cost of PwRC materials is somewhat higher than the usual new materials, the client should consider the premium cost in budget planning:

'...they will also deal with the specification issue around the spec says explicitly not to use this type of thing. So if they can, that would be the can be overcome predominantly. You know, maybe they need to be prepared to pay for it as well.' [C_2P_2]

The participant from the design team C_2P_3 indicated two strategies to overcome the challenges they faced during the design stage of this project. These included developing demonstration projects in which PwRC are used and testing the properties of the material before specifying them to the project:

'...just having a precedent set from this project about what types of applications you can use for, the recycled materials is one...they can be a bit of a risk in being the first project to do something in that way,

> because it can be a little bit untested…And then you may see sort of poor outcomes in another project, if they haven't been installed, properly or selected appropriately. So, materials such as plastic wood will need to be considered in terms of how UV resistant, they are, how they are dealt with, I guess impacts loadings, fire as well.' [C_2P_3]

The participant from the PwRC supplying organisation, C_2P_4, suggested some strategies to deal with challenges that are faced when using PwRC in these kinds of projects. These included support from authorities, revision of material specifications in favour of using PwRC, having supportive individuals on the project team and studying the application of emerging materials. Supportive authorities and up-to-date specifications could significantly remove many barriers to using PwRC in the project because according to this participant:

> '…the specifications are probably the key thing if the specifications allow for the use of recycled material and that takes away a huge barrier and makes it far easier to get acceptance for it in the market…So you know, it's up to the regulators and Vic Roads and Major Projects and people like that to really step up and make sure that there aren't these ridiculous invisible barriers in the way of people taking up the use of recycled material.' [C_2P_4]

> '[and] then there'll be emerging materials that come onto the market, which we'll need to find. Place to use as well. And we need to be open to those sorts of things and be agile and responsive with specifications so that we can broaden the scope of recycled materials and construction.' [C_2P_4]

Chapter 5

Tonkin Gap and Associated Works Project (Case Study 3)

5.1. Project Background

Beginning in 2020, this project, which was led by MainRoads WA, is part of an infrastructure package announced in 2019 and is located approximately 8 km east of Perth, along Tonkin Highway between Collier Road, Bayswater and Dunreath Drive, Redcliffe. Connecting Gateway WA and NorthLink WA, this section of Tonkin Highway forms part of a vital freight and commuter access route known as the Perth to Darwin National Highway. Insufficient capacity to meet current demand at this location has created a bottleneck that constrains the full realisation of the benefits from recent investments in Tonkin Highway and planned future investments to upgrade intersections and extend Tonkin Highway to the south.

Therefore, the Tonkin Gap and Associated Works Project was initiated to address the current bottleneck where the highway was reduced from three lanes to two lanes in Bayswater and Redcliffe, significantly reducing travel times and improving safety. The project is set to also make improvements to access and connectivity for road users, pedestrians and cyclists in the project area. It will provide rail-enabling works for the METRONET Morley-Ellenbrook Line along Tonkin Highway, including underpasses and dive structures, to enable trains to enter and travel along the median of Tonkin Highway and then exit in Malaga. The construction budget for TPG is $400 million, and the completion date is late 2023 (Figure 22).

The head-contractor of this project is Tonkin Gap Alliance (Perth, Australia), the design of the project was done by GHD and the main supplier of PwRC to be used in the project is WA Recycling. The alliance, comprising BMD, Georgiou Group, WA Limestone, BG&E and GHD, is working with MainRoads and the Public Transport Authority (PTA) to deliver the project. Works on this project have been fast-tracked to support WA's COVID-19 economic recovery. In terms of the project timeline, the planning and development were done in 2019–2020, followed by the detailed design that was completed in 2020–2021, early works which began in October 2021, and actual construction started in January 2021. The project is estimated to be completed in late 2023.

Source: MainRoads Western Australia [67].

Figure 22. Construction of Tonkin Gap and Associated Works.

5.2. Project Sustainability and Recognition

This project provides several economic, social and environmental benefits. From the social point of view, the Tonkin Gap and Associated Work Project is expected to create more than 3,000 jobs. It is anticipated that the project would save 17,000 road users 6 and 11 minutes during the morning and afternoon peaks, respectively.

For the design and construction stages of this project, the project's owners used the ISC V2.0 rating system (silver badge) to meet their commitment to delivering positive sustainable outcomes. The project managing team has developed a sustainability policy [68] to reflect their commitment to addressing sustainability issues including growth and economic benefit, efficient resource use, sustainable procurement, and community and stakeholder influence.

5.3. Application of PwRC in the Case Project

The project is using significant quantities of PwRC including recycled concrete, sand and fill material, steel, crushed limestone and aggregate [69]. During the execution phase, the project adopted initiatives that reflect best practice applications to reduce the quantity of virgin materials and remove outputs from waste streams and PwRC where possible. According to the project constructor, these initiatives include the CE, industrial ecology, industrial symbiosis, cleaner production and resource recovery concepts. The following initiatives were adopted within the project:

- Cut to fill optimisation to reduce the import of fill materials
- Use of imported embankment material from another MainRoads project
- Industry engagement to source materials with lower carbon or higher recycled content

The project team uses Roads to Reuse Product Specification – Recycled Road Base and Recycled Drainage Rock [63] to evaluate the characteristics of the PwRC used in this project. As of July 2022, the project had used 430 kt of PwRC including the application of 27.3 kt of crushed concrete [69]. The breakdown of the applications is provided below:

- The reuse of 296 kt of sand
- The reuse of 105 kt of spoil (treated ASS)
- The use of 27 kt of CRC
- The use of 1.2 kt of reclaimed asphalt pavement

5.4. Descriptive Findings: Participants' Profile

The interview process took place between May 2022 and January 2023. The full description of the research approach employed is provided in Chapter 2. Table 9 shows the professional profiles of the research participants in this case study.

Table 9. Summary of participants' profiles in Case Study 3

Stakeholder type	Experience/expertise	Participant code
Client	Experienced sustainability advisor who oversaw individuals from the client organisation who were involved in the project with a particular focus on driving the uptake of PwRC	C_3P_1
Head-contractor	Four years' working experience in the building and construction isector with a focus on major road infrastructure projects in the organisation	C_3P_2
Structural designer	18 years' working experience in the organisation and was involved in the project as the technical director and oversaw the structural design of the work	C_3P_3
Supplier	Working in the organisation as the director since its establishment 10 years ago	C_3P_4

The client's participant (*C_3P1*) was the sustainability advisor in a public agency that was responsible for managing the state road infrastructures. This participant had been involved in many trial projects related to the use of PwRC including CRC in road infrastructures in WA. Their involvement in this project ranged from overseeing the individuals working for the client, to driving ISC rating tool requirements to report the sustainability performance of the project. The participant representing the head-contractor (C_3P_2) had four years' experience of working on infrastructure construction projects in WA. Their role in this project was managing the construction activities in the Tonkin Gap portion of the whole road infrastructure upgrade. They were particularly responsible for civil construction activities. The designer of the project, C_3P_3, worked as a technical director at the design organisation and had worked for 18 years in this organisation in different roles including pavement design manager. Prior to joining the organisation, they had worked in South Africa as a director of a consulting company working with contractors on road projects. The participant representing the supplier of PwRC, C_3P_4, was the director of the recycling facility and had operated the facility for more than 10 years. This participant provided written responses to the interview questions.

5.5. Organisations' Involvement in the Project

The client organisation is a public agency responsible for managing WA state road infrastructure (Table 10). The organisation's involvement includes project planning, setting up the project's alliance, driving the sustainability aspect of the project and superintending the use of PwRC in the project. The design organisation that was responsible for designing the pavement of the project worked with the alliance from the tender stage up to execution. The organisation explored the PwRC applications in the project in close collaboration with the alliance contractors and project management team.

Table 10. Main involvement areas of study organisations in Case Study 3

Organisation	Involvement/activities
Client	Project planning, setting up the project's alliance and driving sustainability (including the increased use of PwRC) within the project lifecycle
Head-contractor	Delivering the project according to the contract, procuring PwRC from the approved facilities and assisting the client to achieve the maximum application of PwRC
Designer	Working with the project alliance team to design the project during the tender and planning stages
Supplier	Supplying PwRC to the project in accordance with WA MainRoads PwRC specifications, and reporting the results of regular waste resources testing to ensure they satisfy the requirements of the specifications

The head-contractor organisation was responsible for delivering the project which also included procurement of PwRC. This organisation aided the client in achieving maximum waste recovery and diversion from landfills through the application of PwRC in the project. The supplier organisation was involved in supplying PwRC with recycled roadbase both grade 1 (permanent pavement) and 2 (temporary roads, access tracks and hardstand areas) materials to the project according to WA MainRoads PwRC specifications. They also were engaged in the transfer and disposal of concrete, sand and asphalt for this project.

5.6. Stakeholders' Interaction Modelling in Case Study 3

Due to the contract being of the alliance type, the companies that participated had a big influence on the project design and delivery including the use of PwRC. According to the client representative, the detailed decisions on using PwRC in the project were made by the alliance and the design organisation, while it was generally supported by the client. However, the client organisation set up the auditing system to ensure the use of PwRC would be safe and according to the relevant specifications. The following model (Figure 23) maps out the interaction between the key stakeholders in this project with a focus on using PwRC.

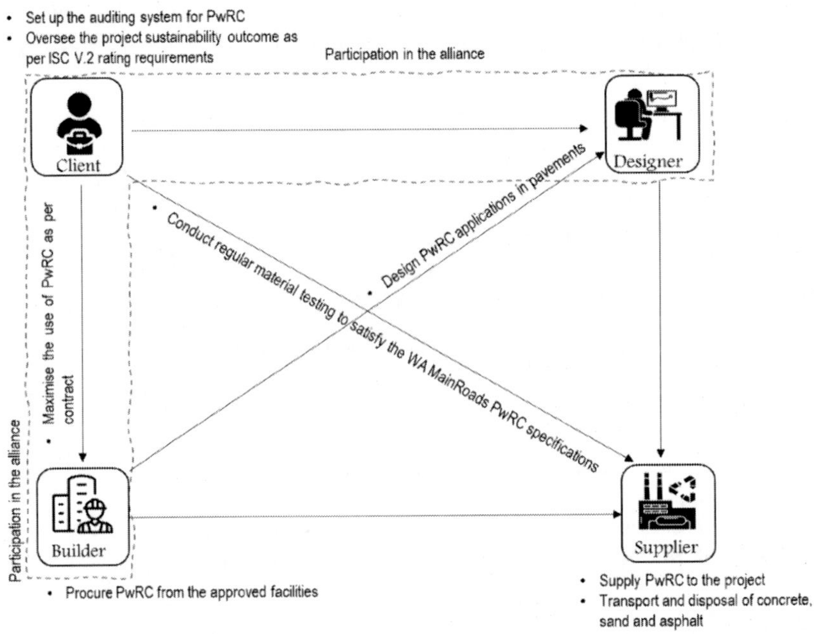

Figure 23. Stakeholders' interaction model in Case Study 3.

5.7. Major Motivations for Using PwRC

Overall, 12 reasons were specified to be the major motivations for using PwRC in this project. These reasons were different in nature but can be grouped under five categories: contract, risk, price, business and social. Figure

24 summarises the responses provided by four participants about their motivations to be involved in this project with a focus on the use of PwRC.

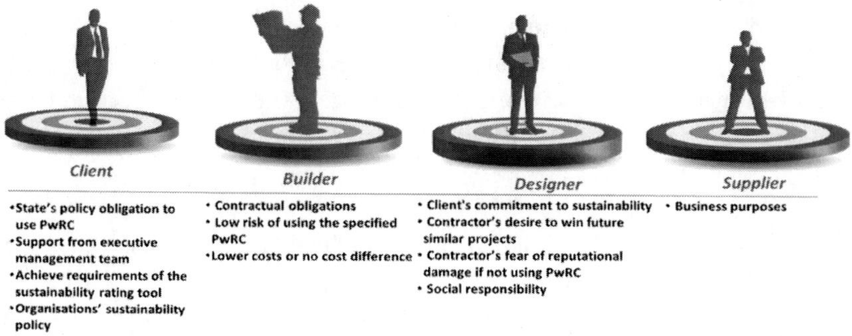

Figure 24. Main motivations for using PwRC in Case Study 3.

The four main motivations for the client organisation (C_3P_1) to use PwRC were 'state-based policy obligations,' 'support from executive management team,' 'achieving sustainability rating requirements,' and adhering to 'organisation's sustainability policy.' The first research participant (*C_3P_1*) indicated that the client organisation has a statutory obligation to use a certain quantity of PwRC in infrastructure projects as stipulated in the state's waste management strategy document [70], which was also echoed in their sustainability organisational policy. They explained:

> 'So we've got obligations with under the West Australian waste strategy to use a certain amount of crushed recycled concrete per year and to drive the uptake within the industry…as well as the MainRoads sustainability policy and using recycled material aligns with that policy.' [C_3P_1]

Furthermore, achieving the credits for using PwRC as stipulated in the ISC rating tool was a motivating factor to use these resources. The organisation's appetite to receive recognition for the sustainability performance of this project was a driving force for supporting the use of PwRC in this project. Hence, there was strong support from the client's executive management team, according to this participant:

> 'So, the project will be looking for different ways of getting points under the [ISC] tool and getting credit. So, therefore, recycled material is one way that they potentially can get credit.' [C_3P_1]

'So, it's not just the project managers or the project directors requesting this to happen, we've had the help of our managers.' [C_3P_1]

According to C_3P_2 the main reasons for using PwRC were the contractual obligations, low risk in using the selected PwRC and the lower cost and the lack of cost differential between these materials (i.e., crushed concrete) and quarried material (i.e., the crushed rock and limestone, respectively). They stated:

'So, there's the contractual obligation. And then there's the cost. So, the cost between crushed recycled concrete and crushed rock road base recycled concrete, it's the cheaper alternative to the cost between CRC and crushed limestone. It's similar. So, there isn't much for saving either way.' [C_3P_2]

C_3P_3, the design manager of the project, mentioned four main reasons for using PwRC in this project. These were the client's strong commitment to sustainability, the contractors' desire to increase their chance of securing future construction projects, the contractor's fear of reputational damage due to not using PwRC as expected and social responsibility. They characterised the scenario as follows:

'You also have to realise that when you win, when you win a contract as an alliance, a lot of the things that motivate the client to pick a certain team is based on their team, their people, and their attitude towards things like sustainability…But there are some more big projects coming. And why wouldn't this team want to bid on that job and be able to put their hand up and say, we use crushed recycled demolition waste?.' [C_3P_3]

'In some contractors' minds, it's, well, why do I need to bother with this client can do nothing about it. But the reputational damage of not doing it was huge…. So if the contractor said I'm not using crushed recycled demolition waste, [the client] will see that in a very bad light, and it's not in the spirit of the contract.' [C_3P_3]

'You know, my, my driver for sustainability is I have a social conscience and I believe it's the right thing to do…I think you'll find a large proportion of the staff in those contractors recognise that as well. I mean, they're all mums and dads they understand and so there is this social conscious that we need to do it.' [C_3P_3]

The last participant in this case study, C_3P_4, indicated that the main motivation for them to be involved in this project was purely on business grounds. As indicated before, the organisation is the supplier of PwRC and a specialist in road waste management.

5.8. Major Barriers to Using PwRC

According to the client's representative (C_3P_1), there were many barriers to using PwRC in this project (Figure 25). Some of the most important ones were 'resistance by other stakeholders,' 'clients' minimum role in decision-making process,' 'complexity of working with alliance,' 'PwRC quality compliance,' 'PwRC performance and insurance,' 'existing commercial arrangement across supply chain,' 'negative perceptions,' 'risk-averse design team,' 'availability of PwRC' and 'contractor capability to use PwRC.' The participant reported that at the beginning of the project, there was resistance by the alliance partners to use PwRC due to the cost and other risks involved. This challenge had a significant impact on the use of these resources initially given that the commercial decisions were made by the alliance core team. They recalled:

> 'We've experienced some resistance from the alliance partners to use recycled materials however, they've come on board.' [C_3P_1]
>
> 'The commercial decision-making is with the alliance partners and not with us as the client. So, to get that decision to be made, has taken quite a bit of effort from our people.' [C_3P_1]

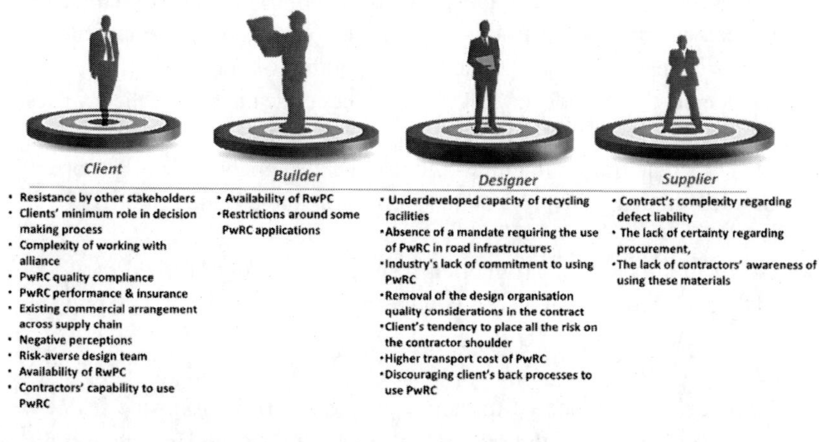

Figure 25. Major issues around using PwRC in Case Study 3.

The nature of the contract necessitates the interaction and close collaboration between different partners involved in the alliance. This means that the decision-making process was subject to many factors emerging from the complexity of the project team's structure. The other two barriers according to C_3P_1 were related to the PwRC quality compliance and their actual performance in this project, which had an enormous impact on the decision on the use of PwRC:

> 'Obviously, within an alliance such as Tonkin Gap, there are different companies all working together, which have all different cultures, and therefore different risk appetites, and then different capabilities that go together with it in terms of driving this outcome. So, I think those competing cultures can have become a challenge of using the product.' [C_3P_1]

> 'Probably the main challenge is making sure that the recycled material you wanted to use is appropriate to be used in its end, from within its purpose on the road as part of the road construction. So, for example, crushed recycled concrete is suitable for use as a subbase under full-death asphalt, it's engineering appropriate. So then with that, it's, you know, you got to make sure that meets our standards or specifications.' [C_3P_1]

The participant mentioned that the use of PwRC in this project was also limited by the alliance partners' negative perceptions, risk-taking behaviour and the contractors' capabilities to use these products:

> 'I think there's a cultural challenge as well around some of the challenges are seen as a risk some and then that relates both to the risk appetite of using a different product, but also the capability of being able to use this product and the confidence that the construction team can use the product and use it properly and handle it properly to get the outcome that we've got'...So that one of them would be the design houses may not be confident in using the product and warranting the product.' [C_3P_1]

> 'Given that there was an internal resistance to using crushed recycled concrete, it's likely that there's commercial arrangements, so that prevent it being used, as well.' [C_3P_1]

Despite the existence of an active waste recovery industry in WA, the participant indicated that the availability of PwRC at the right time, with the right price and for supply to the project in the required quantity was a challenge:

> 'Another commercial challenge was around getting enough product at the right time at the right cost. So once you've proven that it's okay to be used, then it'll be can we get it when we need it, and get it for a reasonable price and it's competitive.' [C_3P_1]

The two major challenges reported by C_3P_2 were the availability of the quantity of crushed concrete required for the scale of the project and restrictions regarding using PwRC in the vicinity of waterways:

> 'The probably the main challenge is the supply. So, because we're using this crush recycled concrete, fairly new product to Main roads into Perth for road building. So, the suppliers typically won't have large stockpiles of material...there's a restrictions around using this material near waterways. So, when you're within X distance of groundwater, you need to switch to limestone or a crushed rock product.' [C_3P_2]

The designer of the project noted seven obstacles associated with the utilisation of PwRC in this project. The first obstacle was the underdeveloped capacity of state recycling facilities to produce the required quantity and quality of PwRC. The second issue was the absence of a mandate requiring the use of PwRC in road infrastructures. The third factor was the industry's lack of commitment to using these alternative materials. The participant explained:

> 'One of the big issues I have at the moment is MainRoads has approved certain suppliers...The reality is they can't produce it [recycled asphalt]. They don't have the asphalt equipment, manufacturer equipment to produce it...So, they market that they have these approved mix designs, but they can't implement it...if you cannot produce a certain product, there's a penalty you have to pay. And I think the asphalt supplier is now paying this penalty.' [C_3P_3]

> 'I think the biggest roadblock is MainRoads were very, very slow in trying to mandate it, they don't mandate it.' [C_3P_3]

The fourth challenge was the removal of the design organisation quality considerations in the contract on the recycler's request, which resulted in noncompliance of PwRC features with the design requirements. The fifth challenge was related to the client's tendency to place all the risk on the contractor's shoulder and not share the risk of using PwRC. Furthermore, the higher transport cost of PwRC due to its physical properties was a major barrier. PwRC has lower bulk density, meaning it requires more space to ship to the site compared to alternative quarried materials. The participant described how this played out practically:

> 'The asphalt supplier said, don't worry, we've got all this covered. We don't need these, this commercial model is in the contract, and they took it out. Now what's meant what's what has now materialised is their product doesn't actually meet our performance requirements in terms of modulus. So we have no stick to bear on them to resolve this' [C_3P_3]

> 'So, there is a hidden, there is an underlying nervousness to do anything that increases the risk on the contractor in terms of pavement performance. But I'm going to lay some blame at [client], they don't necessarily come to the party, when it comes to innovation' [C_3P_3]

> 'The loose bulk density of crushed recycled demolition waste, is let's say 1,800 kg/m3, that's loose. As a subbase [the alternative] is crushed limestone, which has a loose bulk density of about, let's say 1,600. So crushed, recycled demolition waste is heavier. So what's happening is the suppliers are pitching the price of the product...but the haulage costs of crushed recycled demolition waste are higher, because a truck can only carry a certain number of tonnes. So, if he's carrying crushed recycled demolition waste, he carries let's say, 10 or 15%, less material. So, he has to have 10 or 15% more truck movements.' [C_3P_3]

Lastly, the participant mentioned that despite the initial client's objective to use PwRC in this project, their processes will discourage further usage of them in the project:

> 'They said they wanted sustainable sourcing of material, it was a massive opportunity. And they didn't do it. So, they say one thing on the one hand, but their systems and their processes, stop them from doing it on the other hand.' [C_3P_3]

The representative of the supplying company (C_3P_4) indicated there were three main clusters of barriers hindering the utilisation of PwRC in this project.

These included the contract complexity, the lack of certainty regarding procurement and the lack of contractors' awareness of using these materials. On that first point, the participant noted that the contract was complex and was not customised for infrastructure projects. One of the main complexities in the contract was the limitation and the period of defect liability set at five years or more for the entire job and beyond. They explained:

> 'Most of the tenders have initially required the tenderer to accept liability to the extent possible, meaning liability for the cost of the entire job and beyond. Practically our material would not cause material defects and commercially this risk cannot be taken on.' [C_3P_4]

> '[extended defect liability is] not commercially viable and once the goods are inspected and handed over, we lose control of them and there are various permutations that can arise from that point all of which we have no control over.' [C_3P_4]

In terms of procurement, according to C_3P_4, the PwRC procurement did not specify the quantity or the time of the materials procurement, and it was done on an ad hoc basis. Furthermore, the participant highlighted that some of the contractors did not have the knowledge required to handle PwRC and education was required:

> 'In relation to the quantity and when the material was required for the project, communication by the Tonkin Gap could be improved more to provide clarity on when and what volumes of material are needed, we need to balance the needs of all our clients when supplying material and we have a finite capacity to produce material…At the time of tender, we receive a bulk volume, but it is not stated at the time of the tender when and how much material is required. It is in our best interest to continuously communicate with our client and receive regular histograms of the material required on a monthly basis. We have had no major issues so far even in the busy times.' [C_3P_4]

5.9. Major Strategies to Overcome the Reported Barriers

In response to major strategies to overcome the barriers reported, participants provided 14 factors (Figure 26). The first participant (C_3P_2) indicated five main strategies that can address the barriers to using PwRC in construction projects. These included 'establishment of audited system,' 'mandating the use of PwRC in contracts,' 'financial incentive,' 'overhaul the existing long-term agreements,' 'positive experience using PwRC' and 'training and education and awareness raising.'

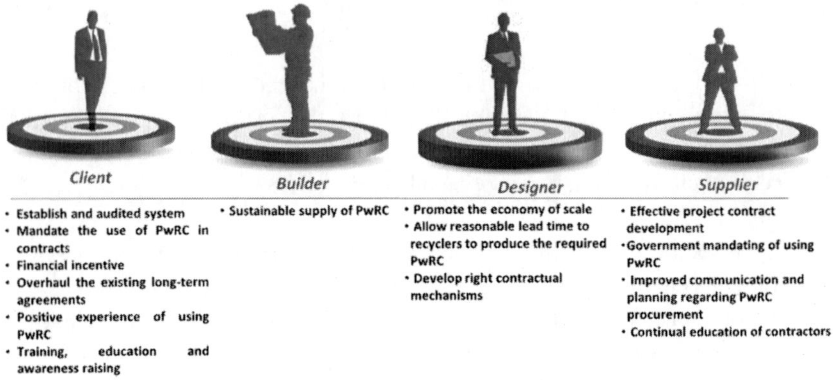

Figure 26. Major strategies to remove barriers to using PwRC in Case Study 3.

C_3P_1 maintained that the first strategy is to have an audit system to ensure that PwRC in use has the quality and performance as required. The second and third strategies represent the carrot and stick approach where the stick was to mandate the use of PwRC in construction contracts and the carrot was to provide financial incentives by paying the right price for applying PwRC in construction projects. They stated:

> 'So, we've done a number of things to overcome the issues around standards and specifications. So we've got a specification that the crushed recycled [concrete] must meet, and then have an audited system to assist with that.' [C_3P_1]

> 'I think to overcome some of the commercial challenges, there has to be that balance between incentives and the mandate of the outcome that you want as a client. So, I actually think mandating the product is probably the way to go in contracts…So you might have an incentive to use the product and have the right price associated with using crushed recycled concrete, to make sure that there's enough incentive for people, just for companies to start to go down in that direction, but you want to set it up.' [C_3P_1]

The participant opined that one thing that needs to change is to overhaul the existing agreements that had been established in the long run. By implementing this strategy, the opportunities for using PwRC will significantly increase:

> 'I believe that there's commercial arrangements internally within companies. So, some companies will have their supply chains have long-term contract agreements with their supply chains. I don't know whether this is the case in Tonkin Gap. But you need to be able to overcome, we're working on a way to overcome those long-term agreements. [but] as a government agency, we want to make sure that everyone's sort of happy.' [C_3P_1]

C_3P_1's last two strategies were related to education and positive experience. Education was particularly emphasised by this participant due to its significant effects on the stakeholders' attitudes, skill development and industry practices:

> 'I think culture comes down to a lot of experience, you need to make sure that there's a positive experience using the product, and make sure that there's enough training and education and awareness that can be delivered to the industry on using the product and what they need to do to use it safely, so it works. And they're confident with those processes and systems and it's got that endorsement. I think, training and education can go a long way in changing culture, and then changing practice on projects.' [C_3P_1]

C_3P_2 suggested that planning for the sustainable production of PwRC in WA is an area of improvement to overcome the current issues. The main reason for this suggestion was that the participant believed that WA MainRoads's plan is to keep using these materials in its road infrastructure projects, which creates problems regarding long-term supply. The three main strategies suggested by the design organisation representative, C_3P_3, to overcome the challenges are classified under four categories: Cost, procurement, contract and collaboration. The strategies were: promoting economy of scale, allowing reasonable lead time for recyclers to produce the required PwRC and developing the right contractual mechanisms. The economy of scale will motivate investment in improving recycling processes, which in turn reduces the PwRC costs and makes the products commercially competitive. As the participant explained:

> 'I think a lot of these issues will be solved by economies of scales...Like recyclers, they're not motivated to get a lot of recycled materials, because they know that the products won't be as popular as the virgin materials. But if like more pro, like more organisations across Australia started, like using this material so that they get more of this stuff. So the whole processing costs will be reduced. And then eventually the price would be kind of competitive, comparable to the virgin materials.' [C_3P_3]

According to C_3P_3, giving the PwRC producer motivation to invest in their processing activities will result in a better outcome, such as high-quality production of high-quality materials that can pass all the requirements and be safely used in construction projects. In particular, by giving enough time and reasonable order to recyclers, the client will be able to create a market for PwRC. This includes the procurement of PwRC from different recyclers to ensure the quantity required will be prepared when needed. This kind of procurement was described as smart procurement by C_3P_3:

> 'If you said to the industry, I need 500,000 tonnes of product, I need it in two years. I recognise you may not be able to produce it all, but I'll pay you for everything you produce. Now, if you're a smart businessman, what are you going to do? You're going out there and finding as much waste as you can. Because the more you process and get onto MainRoads property, the more money you make. That's where the investment comes.' [C_3P_3]

The right contractual mechanisms will change the industry's behaviour and encourage them to procure more PwRC. Furthermore, the collaboration between project partners to strategically use road demolition resources was described as an important enabler:

> 'So I think if you if you put the right contractual mechanisms in your subcontract, you will drive those behaviours and you And if the contractor if the asphalt supplier develops a good solid market, making mix with 20%, wrapping it, what he will do is he'll be running around the country trying to pick up as much old asphalt as he can. Because he will make money, you'll make more money. And so, it's the right thing to do.' [C_3P_3]

> 'The contractor tendered on the premise that he would win this material [waste materials]. And as a result of that, that's the only way he can

afford to build the road is to be able to reuse these materials. But what he hadn't properly thought through how he was going to win the materials so that the quality was good enough. And we certainly got pretty involved in working out a way to win the material without it degrading. So the enabler here was me and my team, working with a contractor to work out how to get this material off the road without spoiling it.' [C_3P_3]

C_3P_4 provided several suggestions to overcome the issues experienced when supplying PwRC to the project. These suggestions included effective project contract development, government mandating of using PwRC, improved communication and planning regarding PwRC procurement, and education. The participant believed that the contract needs to be customised for road building materials rather than having a one size fits all contract for the supply to a major project. The main strategies under this category included timely inspect of PwRC delivered on-site and annual review of cost rates:

'Once delivered to the site [recycled materials] need to be inspected within a short period of time to avoid issues relating to chain of custody passing and liability relating to the material.' [C_3P_4]

'Generally, the projects that require CRC [crushed recycled concrete] extend over 2 to 3 years. Rates for products supplied need a fair basis to be reviewed annually or more often in certain circumstances (i.e., fuel price rises, transport price rises etc…).' [C_3P_4]

C_3P_4 also highlighted the role of government in increased uptake of PwRC via mandating their utilisation first as a substitute for virgin materials. Better communications and planning in relation to the quantity of PwRC required and when they are required was reported to be an important step in removing the barriers, which will help the PwRC producer to better manage their production. They suggested:

'In relation to the quantity and when material was required for the project, communication by the Tonkin Gap could be improved more to provide clarity on when and what volumes of material are needed, we need to balance the needs of all our clients when supplying material and we have a finite capacity to produce material.' [C_3P_4]

Lastly, continual education was indicated by the participant to be a necessity when a project is using PwRC.

Chapter 6

OneOneFive Hamilton Hill Project (Case Study 4)

6.1. Project Background

This project, which is managed by WA Development, aimed to redevelop an old high school and neighbouring land (11.9 ha) into a residential estate to provide residential opportunities for WA communities (Figure 27). The old school was closed in 2017 after 55 years of operation, according to the WA Department of Education [71]. The project budget was $2.7 m and completed over 12 months in 2019. In this project, through the extraction of demolition resources from the old senior high school, around 86% of waste diversion was achieved.

6.2. Project Sustainability and Recognition

As one of Development WA's Innovation Through Demonstration projects, OneOneFive Hamilton Hill targets the highest levels of sustainability and delivered an ecologically sensitive development—minimising water and energy consumption, while retaining the site's landform, trees and biodiversity.

At the time of the case study, the project had been awarded six petals of the Urban Development Institute of Australia (UDIA) under its benchmark EnviroDevelopment certification program. The petals include Ecosystems, Waste, Materials Energy, Water and Community. Furthermore, the project has been recognised by Water Corporation WA as a Waterwise Platinum development, the highest Waterwise Development status, and also won the Waste Sorted Award presented by Waste Authority WA in 2021. Figure 28 illustrates the main project certifications and awards achieved to date.

The project has delivered some key economic, social and environmental benefits. The estimated cost saving was around $100,000, which was realised in reduced waste levies and transportation costs. This cost-saving could offset the increased cost of completing the crushing process and keep recovered materials suitably stored and organised.

The new estate features approximately 232 lots and offers a diverse range of new housing options, including single residential homesites, town homes and compact lots along with four grouped housing sites. Furthermore, the project reserves the site heritage value by incorporating the extracted materials into the new construction.

Source: TABEC (2022) [72].

Figure 27. Construction of OneOneFive Hamilton Hill project.

Figure 28. Project sustainability accreditations and awards achieved to date.

Lastly, with the minimisation of the need for material movements, the project had a minimum impact on the local road networks. There were 1,000 fewer truck movements, 9,000 L of diesel saved, and therefore, 663 t of Co_2 were avoided. This carbon emission reduction were achieved in three key areas: onsite, through the replacement of raw materials with recovered demolition products; offsite, through recycling; and during transport, through reduced truck movements.

6.3. Application of PwRC in the Case Project

About 10,000 t of waste resources were extracted from the old school using the methodical deconstruction method. Most of the PwRC applications were in the civil and landscape construction works including the drainage system, local access roads, landscapes and footpaths. About 86% of demolition resources were crushed onsite for reuse in the project. Through the reuse of these materials, carbon emissions were achieved by reducing the need for material transportation to and from recycling facilities. The following represent the PwRC quantity and type that were used in this project:

- The use of salvaged timber in landscaping features, e.g., shade structures and seating
- The reuse of 40,000 clay bricks and roof tiles as aggregates underneath the drainage infrastructure
- The reuse of old bricks to create brick walls and a brick toilet block

- The reuse of crushed brick, tiles, concrete etc into the road sub-base replacing the need for newly mined materials like crushed limestone
- The use of 2,425 m^3 of recycled concrete in retaining walls replacing the need for newly mined limestone
- The use of 400 t of PwRC in different applications including constructing temporary truck access roads

Furthermore, the head-contractor has imported 1,050 t of recycled roadbase from other demolition sites to use in this project. The project used IPWEA/WALGA Specification for the Supply of Recycled Road Base [63] to assess PwRC quality and performance.

6.4. Descriptive Findings: Participants' Profile

In order to capture the industry experience in using PwRC in the case construction project a series of interviews were conducted with the representatives of the key stakeholer groups. The full description of the research approach employed is provided in Chapter 2. Table 11 summarises the research participants' experience and involvement in construction projects. The client representative of this project was a senior development manager in WA with more than 30 years' of experience working in the property industry. At the time of the interview, C_4P_1 had worked for 15 years as a development manager in the organisation, primarily on residential projects throughout WA. The second participant in this case study was a civil engineer and the director of a contracting engineering company. This participant is an expert in land development engineering, particularly those related to housing estates, and was directly involved in the design and civil works of the project.

Table 11. Summary of participants' profiles in Case Study 4

Stakeholder type	Experience/Involvement	Participant code
Client	A senior development manager involved in the property industry for more than 30 years	C_4P_1
Head-contractor	A civil engineer and the director of the organisation with 20 years' working experience in the organisation	C_4P_2
Design consultant	The director of a private company specialises in landscape architects, urban design and sustainability consultancy	C_4P_3
Supplier	The director of the organisation with more than 27 years' experience in waste recovery in WA	C_4P_4

The third participant, C_4P_3, had different roles at the time of the interview. In addition to being the director of a private company operating in the field of sustainable design of outdoor spaces, this participant has worked as a senior academic at a WA university with a focus on environmental engineering and provided advice to government agencies including WA Waste Authority. The last participant in this case study (C_4P_4) was the director of a consulting company that specialises in C&D waste management and demolition. This participant's experience in this field included 28 years working on a variety of demolition projects as well as managing several landfills in the state.

6.5. Organisations' Involvement in the Project

In this section, the engagement and the roles of the four organisations involved in this project are explored. The organisation that was responsible for land development and civil works was a consultancy company based in Perth. As the contract administrator, this organisation collaborated with the key stakeholders (i.e., planner, developer, local council and demolition contractor) to achieve the project objectives. In particular, the organisation superintended the demolition contractor to make sure that material resources extracted from the old school buildings could be re-used. The summary of the organisation's roles and responsibilities is presented in Table 12.

Table 12. Main involvement areas of study organisations in Case Study 4

Organisation	Involvement/Activities
Client	Project managing the property redevelopment from its acquisition of the land following the planning process, the demolition, the remediation, the bulk earthworks civil works, and land sales and building phase
Head-contractor	Land development, contract administration, design of civil works, supervising the demolition contractor
Design consultant	Multiple roles on the project team including sustainability consultancy, landscape architect, urban water consultancy, and estate architect on behalf of Development WA
Supplier	Involved in the capacity of demolition contractor and also the producer of PwRC on-site

The design consultancy organisation was appointed to the project at the very beginning, commencing master planning at the time when the school was still operational. This organisation guided the project team and the client on sustainability initiatives and oversaw the process of securing sustainability accreditation for the project. The supplying organisation, which also managed the demolition activities, were engaged through a tendering process and was responsible for demolition and asbestos removal. However, a few weeks into the operation, they were told by the project team to have a particular focus on processing and reusing the demolition resources.

6.6. Stakeholders' Interaction Modelling in Case Study 4

Figure 29 shows the interaction model between the four study stakeholders in Case Study 4. The modelling was done based on the responses captured during the interviews.

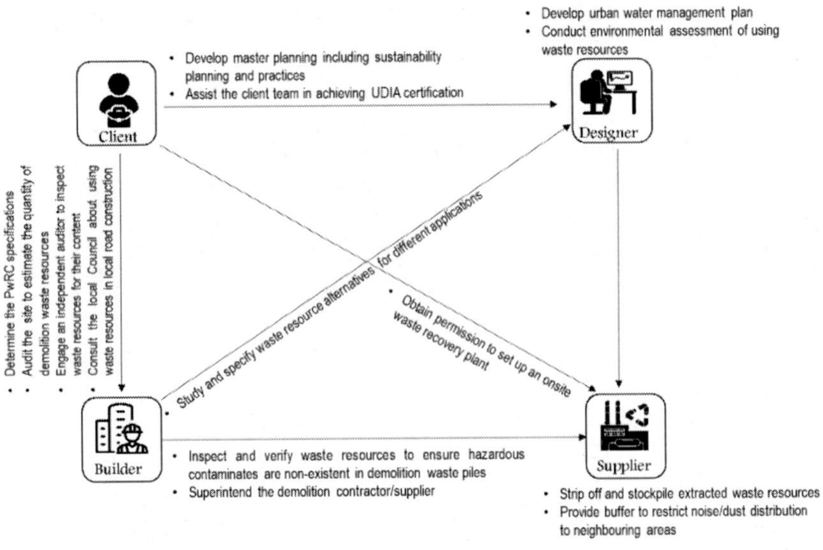

Figure 29. Stakeholders' interaction model in Case Study 4.

6.7. Major Motivations for Using PwRC

As illustrated in Figure 30, the participants in this case study highlighted various reasons for using PwRC on the construction site. The most frequently mentioned factor was to reduce waste disposal or otherwise increase the waste recovery rate.

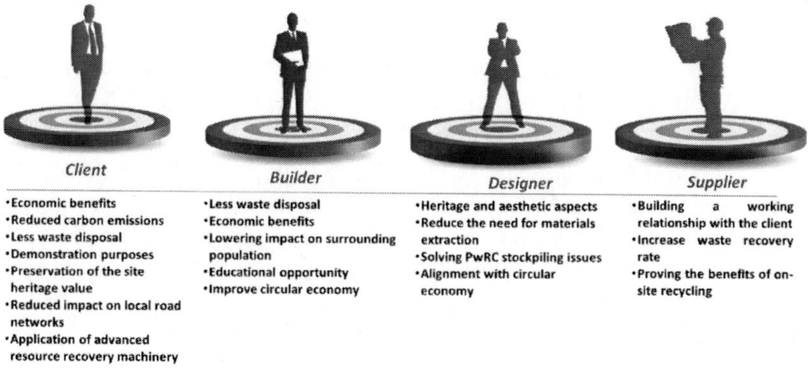

Figure 30. Main motivations for using PwRC in Case Study 4.

C_4P_1 gave various justifications for the utilisation of the PwRC in this construction project. The reasons included economic benefits, reduced carbon emissions, less waste resource disposal, demonstration purposes, preservation of the land's heritage characteristics, reduced impact on the local road network and the application of modern waste recovery technologies. As previously mentioned, the project utilised a substantial amount of waste resources on-site, resulting in cost and carbon savings. This participant stated that their unique waste management solution was deemed superior to landfill disposal and obtaining PwRC from recyclers. Furthermore, the use of PwRC reduced the need for material transportation leading to minimum impact on the local road networks. They explained:

> 'So, they actually realised that they could save some money by not tracking all that material away. They realised that the cost to crush it on-site was less than tracking the material off-site...Because this purely worked out to be the most economical, most advantageous in terms of carbon footprint and disruption to the residence.' [C_4P_1]

> 'We put recycled aggregates from the school under the drainage infrastructure, so we didn't have to buy and import those new materials. And that way, we're basically cutting down on what new material needs to be mined.' [C_4P_1]

> 'The truck movements and the impact on the road networks, the impact on the suburban residential streets, the impact to people, through truck movements and the dangers, you know that is far worse than allowing a temporary facility that's going to crush their own material on-site.' [C_4P_1]

Furthermore, this project was an opportunity for the client's organisation, which is a state government agency, to demonstrate the successful utilisation of PwRC in residential construction to other public and private stakeholders:

> 'Certainly, the purpose of these projects for us is to be able to demonstrate industry. Whereas as state government, obviously, we want to be showing that we're pushing the boundaries and trying to, I suppose provide demonstration projects that show what can be done and within a normal commercial environment.' [C_4P_1]

Lastly, the land on which this project was being built had some heritage value in two periods: pre-colonisation indigenous and European. The indigenous population used the land as scrubland and bushland as well as an important place to survey their surroundings due to its location on a hill. In the European period, the land was created, farmed and then became a school. According to C_4P_1, the use of PwRC derived from the old school could preserve the heritage spirit of the residential development:

> 'In terms of the school heritage, we actually retained a lot of material that we didn't crush that we actually reused in landscaping elements and for different features within the state that then show that sort of that old school being reconstituted into the new development...we're also providing that opportunity for people that perhaps went to this school to come back and look at the new residential development and still recognise certain elements out of the school that they might have gone to landfill.' [C_4P_1]

The head-contractor representative, C_4P_2, highlighted the six motives for using PwRC in this case project. The diversion of waste resources from landfills and the cost savings associated with it were the first two main reasons why their organisation was motivated to use PwRC. They described:

> 'We then had huge stockpiles all over the site of crushed materials and ordinarily, they would have been put into trucks and disposed of quite expensive to a landfill facility that would have been ordinary. Probably outcome we thought, well let's assist the opportunity of reusing this as recycled Rd material and there was probably that opportunity to promote the circular economy.' [C_4P_2]

Furthermore, C_4P_2 indicated that the use of PwRC primarily extracted from the demolition resource reduced the impact of construction activities on the surrounding population. By minimising the truck movements due to keeping demolition resources on the site and subsequent applications, and not having to ship new materials to the site, the usage of PwRC provided positive social and environmental impacts. The educational opportunity in using PwRC was the last motivation indicated by C_4P_2. The participant stated that PwRC was an opportunity to educate their organisation, design consultants and the wider industry about the possibilities of using PwRC in construction projects. Lastly, C_4P_2 mentioned that the client's strong support facilitated the usage of PwRC in this project. They explained:

> '...[we wanted to] educate consultants and the wider industry on what can be done, how you can do it and try and, you know educate ourselves that we can take that kind of projects and slowly sort of do our little bit in our pushing the industry forward and sharing that knowledge you know throughout.' [C_4P_2]

> '...[the client was] very supportive in that respect that you don't just sort of limit you know education to one project or one that a group of people try and do your part in nudging the whole industry forward in that respect so. And there was actually consistent support from [the client] the whole along there.' [C_4P_2]

In response to the reasons why the design consultancy organisation was involved in this project and pursued the utilisation of PwRC, C_4P_3 indicated that environmental impacts, heritage value and aesthetic aspects of PwRC, were the two main drivers. In terms of environmental sustainability impacts, the participant noted that the utilisation of PwRC reduces the need for new extractions, solving the ongoing issue of stockpiling in the state and alignment with the CE aspiration:

> '...in line with a circular economy aspiration, recognising that there is value in these in these materials, you know, reduce the need for the extraction and processing of new materials, and avoiding the issue of seeing materials coming out of the demolition of this project stockpile which we know is an ongoing.' [C_4P_3]

The two last motivating factors were related to the heritage value of the construction site, which the participant thought should be reflected in the new design. Furthermore, the PwRC produced from the demolition resource had strong visual impacts that were perceived to be beneficial from a design perspective. They explained:

> 'It was very clear from the early community consultation with the public, that people wanted a sense of connection with the old school. So they didn't want to be erased…you know, reusing the old bricks, reusing salvaged timbers reusing some of the other school elements in public art, and so forth, is a way of, of keeping that character.' [C_4P_3]

According to C_4P_3, for the demolition and recycling contractor, the three motives for being engaged in this project were building a working relationship with the client, increasing the waste recovery rate and proving the benefits of on-site recycling:

> 'For me. It was purely and simply about getting maximum recycling and reuse on-site and working with the client WA Development who was sensational and working with them. So it was really clear that we had to maximise reuse and recycling. That was really my goal.' [C_4P_4]

6.8. Major Barriers to Using PwRC

The participants in this case study mentioned 15 unique challenges that hindered the application of PwRC in the OneOneFive Hamilton Hill project. As illustrated in Figure 31, the most frequently mentioned challenge was the delays that occurred in getting permission to use the temporary waste recovery plan on the demolition site.

The client's respondent, C_4P_1, noted four major barriers to using PwRC in this project. First, the unsupportive state waste regulations made the permission for temporary setup of waste recovery (e.g., crushing) machinery unreasonably long. The participant mentioned that they had to go through the same process and meet the same requirements as they would in the case of establishing a permanent recycling facility. They recalled:

> 'I think at least three months, maybe four months for approval, we had to meet all the same requirements as a permanent facility on the noise side of things, which actually doesn't make sense…It effectively meant we had to go through a process if we were setting up a permanent crushing

facility that was going to be operating for, you know, 5, 10 years or something. So that process is something we tried to challenge and we were unsuccessful because it is legislated.' [C_4P_1]

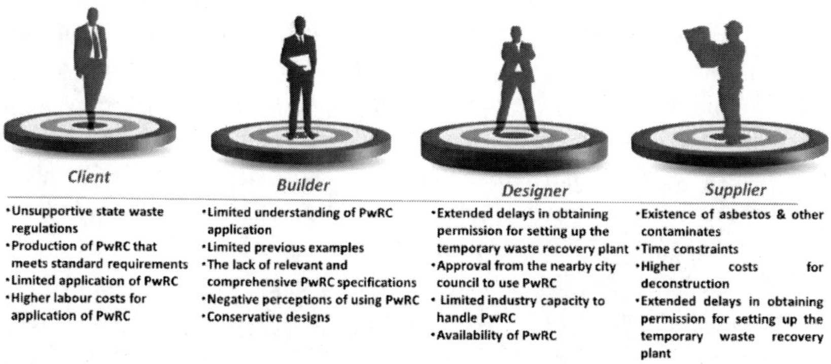

Figure 31. Major issues around using PwRC in Case Study 4.

The second challenge reported by C_4P_1 was to ensure that the final PwRC would meet the standard requirements for the intended applications. For this to occur, the participants had issues fully separating the waste resources on the site:

> 'In that [Engineers Australia] specifications, there's certain percentages that are allowed. So that was a big challenge for us how the demolition contractor was separating bricks from concrete. Sometimes they're attached, obviously. So, if you're talking [about a] big high school, where there's concrete and columns, bricks built in between the columns, then typically the material just all comes down together.' [C_4P_1]

The two last challenges were related to the application of PwRC. The participant observed that PwRC cannot be applied everywhere due to regulatory restrictions or lower performance. Moreover, there was a higher labour cost for using salvaged materials compared to brand-new ones as it needs different handling and application methods. They explained:

> 'So really, what it became a challenge there just to try and find what at every opportunity, where could we put recycled material versus new material. I guess that's what we've tried to do. It's not always possible.' [C_4P_1]

'There's probably a cost involved because buying new bricks is cheaper than the labour required to say chip off the old mortar off an old brick.' [C_4P_1]

The second participant, C_4P_2, who represented the organisation overseeing the civil works named five barriers: the limited understanding of PwRC applications, limited previous examples, the lack of relevant and comprehensive PwRC specifications, negative perceptions of using PwRC and conservative designs. C_4P_2 noted that the application of PwRC was not common and there were very limited examples of this application that could be used as a reference. Furthermore, the participant mentioned that the only available specification [73] could not help the design team understand the best applications for the available PwRC in the project. They stated:

'There was not a lot of sort of strong understanding because there had been in sort of to my knowledge only limited examples of recycled C&D waste being used in road construction in Perth.' [C_4P_2]

'There was no common specification that we could go and refer to or point to and say oh that's appropriate to use on this project. There's a WALGA specifications…So that's what I guess sort of specifications with a no clear reference that we consider turn to and so what we did is essentially developed our own sort of bespoke specifications.' [C_4P_2]

C_4P_2, highlighted the negative perceptions from the local council regarding the use of PwRC in road construction projects. The participant indicated that a previous application had not performed well, which created negative perceptions among stakeholders. This in turn was a challenge for design works, resulting in the production of conservative designs:

'There was some negative perception with the local council who sort of take some ownership of the roads, be due to sort of, I guess, historic trials I say of roads that had been constructed using recycled materials, and the asphalt had done performed poorly over time, and so they wanted to know how that was sort of being addressed.' [C_4P_2]

The design organisation's representative, C_4P_3, noted five primary challenges in using PwRC in this project. These include extended delays in receiving permission for the temporary waste recovery plant operation on the site, approval from the nearby city council to use PwRC, the industry capacity to handle PwRC and the availability of PwRC. The latter factor caused several stages of design refinements to address the issue of PwRC availability:

> 'In terms of the design and detail for the landscape architects, again, no problems with the contractors being happy to use the materials. But just at times, you know, when designing a large project, but only having a certain amount of materials, that it just means you have to be more nimble in how you're designing.' [C_4P_3]

C_4P_4 noted several challenges in using demolition resources, most prominently, the existence of asbestos and other contaminants in the waste resources:

> 'So, the biggest challenge was asbestos. We found asbestos in all sorts of places randomly in the buildings behind whiteboards behind blackboards behind cupboards in window frames underneath and packing them out…and so that was the biggest challenge was making sure the material we crushed was clean.' [C_4P_4]

The other challenges included time, cost and getting permission for waste recovery on-site. In terms of time, the client sought to complete the project without delay. According to C_4P_4, there were extra costs for additional work to ensure that the waste resources extracted from the old school were free of contaminants. The extra work also impacted their organisation's operation on site:

> 'The time on the asbestos removal and I think in the future you would just change your methodology. Because obviously, we were finding it ad hoc. So, we'd start another building...So while we're doing that, we have to go find something else to do. So, it was a bit all over the place because of that issue.' [C_4P_4]

> 'Then cost, because of the extra work that we had to go through to make sure that all asbestos was collected and removed. Obviously, this cost impacts the job in doing that.' [C_4P_4]

> 'We're in the middle of a residential area and we're talking about crushing. You're basically in Perth. It never happened in a residential area before. So that took, I think, almost six months to get approvals to be allowed to crush.' [C_4P_4]

6.9. Major Strategies to Overcome the Reported Barriers

The participants in this case study provided 12 strategies that could potentially overcome the issues faced in this project. The main strategy that was indicated by three participants was to set up demonstration projects in which PwRC are successfully used. Figure 32 summarises the key findings in this section.

Figure 32. Major strategies to remove barriers to using PwRC in Case Study 4.

The participant representing the client in this project provided four main strategies to increase the utilisation of PwRC and also remove the barriers mentioned. According to C_4P_1, the local councils should come on board at the early stage of the project development, as some PwRC can be used in their own infrastructure projects:

> 'So when you're constructing those roads, you have to construct them in a way that's acceptable to the council because they have the long-term liability to maintain those roads...So, it's important that the council comes on board because councils also manage footpaths.' [C_4P_1]

The second was to revise the regulations determining the permission for establishing waste recovery facilities. The participant suggested that permission should be given on a project-by-project basis and revised in favour of on-site waste recycling and production of PwRC due to the massive benefits associated:

> 'I think put into the minds of the regulators that they need to probably consider a different process for short term operations to affect that we could crush the whole school and around about three weeks…But it has to be done on a site-by-site and project-by-project basis, it's not going to suit every demolition…I suspect that if we hadn't crushed this material

on site, the amount of recycled content in our project would be lower than it is now.' [C_4P_1]

Furthermore, studying the variety of applications for PwRC can increase the uptake of these materials. In this project, according to the client participant, the project team workshopped the opportunities for PwRC applications. Even though they cannot be used in all circumstances, the project team was successful in finding some innovative applications for these materials:

'What we ended up doing was just doing a blanket instruction to all consultants that in all their design work, whether it's the civil engineer or the landscape architect, or anyone that's involved in the land development phase of the project, is that where there's a recycled material option to specify that…we basically said, look, brief to us that if [it] can be built in recycled material at the same cost.' [C_4P_1].

The last strategy was to facilitate the demonstration aspects of construction projects in which PwRC are used. The client organisation used this strategy to promote this idea among the major stakeholders in WA:

'When we do the projects, we're now looking at ways to demonstrate. So we had a tour of the project…and we talked through one of Perth, sort of bigger private developers, their project managers came through so we're able to point out to them, the recycled material, the fact that we'd had these stockpiles of crushed material, showed them our retaining walls and asked them if they recognise them and they didn't recognise them.' [C_4P_1]

The builder representative, C_4P_2, reported that having demonstration projects, education on design considerations and local knowledge are the main strategies to promote the use of PwRC. They emphasised that:

'Probably [there] needs to be an educational piece around what are the design considerations that you need to be sort of thought through…before you might sort of charge off and say I wanna do a project with recycled base. OK let's step through what you need to do there.' [C_4P_2]

C_4P_3 offered three strategies to reduce the impact of challenges towards using PwRC in construction projects. The strategies include effective time management planning, early engagement of stakeholders and mandatory site visits for contractors. According to the participant, better time management can reduce the impacts of different risk factors and will provide insight into the time implications. Likewise, they noted that early engagement of stakeholders will contribute to managing risks and ensure that all parties involved understand the project goals. They explained:

> 'It's about good planning, understanding the timeline implications. Certainly, [through time management planning] you would hope that early engagement with regulators helps identify risk and ensuring that all parties understand what the intention of the project is. So for example, once we designed the landscape and packaged it up, we had a section in the landscape specification that actually articulated the intent of waste avoidance and resource recovery for the project.' [C_4P_3]

According to C_4P_3, a mandatory site inspection is a common practice in the industry to make sure that the pricing of the job is done with a high level of confidence and that the project goals are clearly understood by the contractors. They explained:

> '...[we] had a mandatory site visit inspection for contractors who are going to be tendering for the landscape construction works. I'm pretty sure the civil engineer did the same for the civil contractors to come and actually see the material to understand what was there and before they did their pricing...So that would improve our confidence that they would be pricing appropriately.' [C_4P_3]

C_4P_4 suggested that having a clear contract that is informed by negotiating sessions would diminish a lot of issues concerning time and money. This participant noted that, for instance, in the negotiations it had to be clear that it is a full strip-out kind of job with the potential for having asbestos:

> 'So it's probably the only way to overcome that is if you're going to do a major crushing and reuse on-site, Is that in the initial contract negotiation that you need to make it clear that it's a full strip out with the potential for asbestos, that would probably save a bit of time and money, potentially.' [C_4P_4]

The other suggestion was to showcase the project to set an example of such activities to ensure that future projects will run more smoothly. The participant thought this would aid in getting the necessary approvals in a timely manner and also in determining the best methodology for handling the waste resources.

Chapter 7

Integrated Analysis of Case Studies

7.1. Introduction

The following sections outline the research findings emerging from the second round of analyses applied to findings from the four case studies (Chapter 3-6). First, in Section 7.2, the industry experience and profile of all 16 research participants are cross tabulated. Second, in Sections 7.3 to 7.5, the motivations, barriers and strategies that were captured in the four case studies are juxtaposed. Third, by aggregating these factors, the major relevant categories are developed. The new categories aim to give an overall picture of issues around and opportunities for using PwRC in the Australian building and construction sector. Last, in Section 7.6, the most frequently observed categories across the case studies and stakeholder groups are compared.

7.2. Study Participants' Work Experience

The following table summarises the profiles of the research participants, including their construction and waste recovery industry experience and knowledge (Table 13). The majority of participants were substantially involved in the case projects, allowing them to provide valuable information on the application of PwRC in the respective examples.

7.3. Integrated Analysis of Motivations for PwRC Application

In response to the question enquiring about the main motivations for using PwRC in the selected case studies, the research participants referred to 64 factors (Figure 33). An analysis of the responses suggested that the four stakeholders provided a fairly similar quantity of motivation factors during the interviews.

Table 13. Summary of participants in the case studies

Case study / Participant	Client	Head-contractor	Designer	Supplier
Case Study 1 (Brickworks shopping centre)	20 years' experience in construction project development	About six years working experience in the organisation	11 years' experience in architectural management	Seven years' working experience in the organisation as the sale manager
Case Study 2 (Mordialloc freeway)	Senior project manager with extensive experience in project managing of public infrastructure projects	20 years' experience in the building and construction sector as a site engineer, project engineer and manager	15 years' experience of working as design consultant	Highly experienced corporate communicator with a background in government, corporate, industry and community organisations, with 4-year employment history in the organisation
Case Study 3 (Tonkin Gap and Associated Works)		Four years' working experience in the building and construction sector with the focus on major road infrastructure projects in the organisation	18 years' working experience in the organisation and was involved in the project as the technical director and oversaw the structural design of the work	Working in the organisation as the director since its establishment in as a director of the organisation for 10 years
Case Study 4 (OneOneFive Hamilton Hill)	A senior development manager involved in the property industry for more than 30 years	A civil engineer and the director of the organisation with 20 years' working experience in the organisation.	The director of a private company specialises in landscape architects, urban design and sustainability consultancy	The director of organisation with more than 27 years' experience in waste recovery in WA

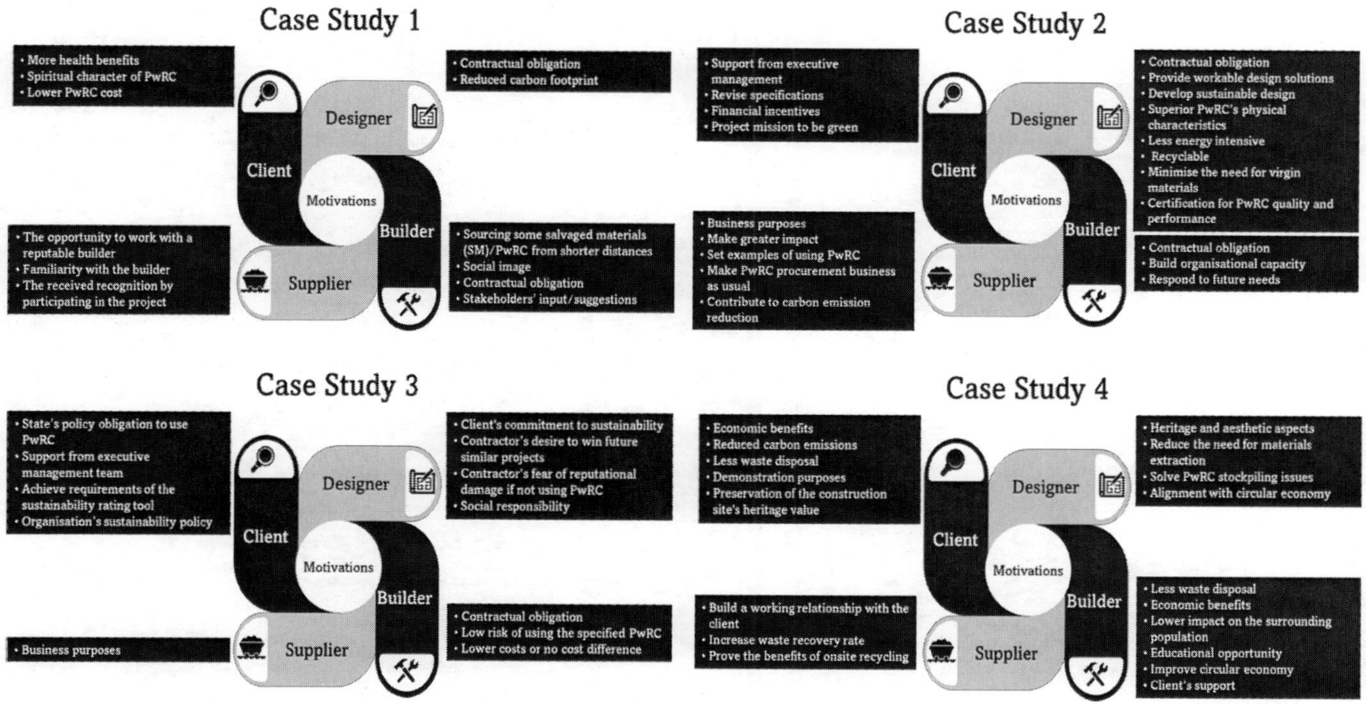

Figure 33. Motivations for using PwRC in the study construction projects.

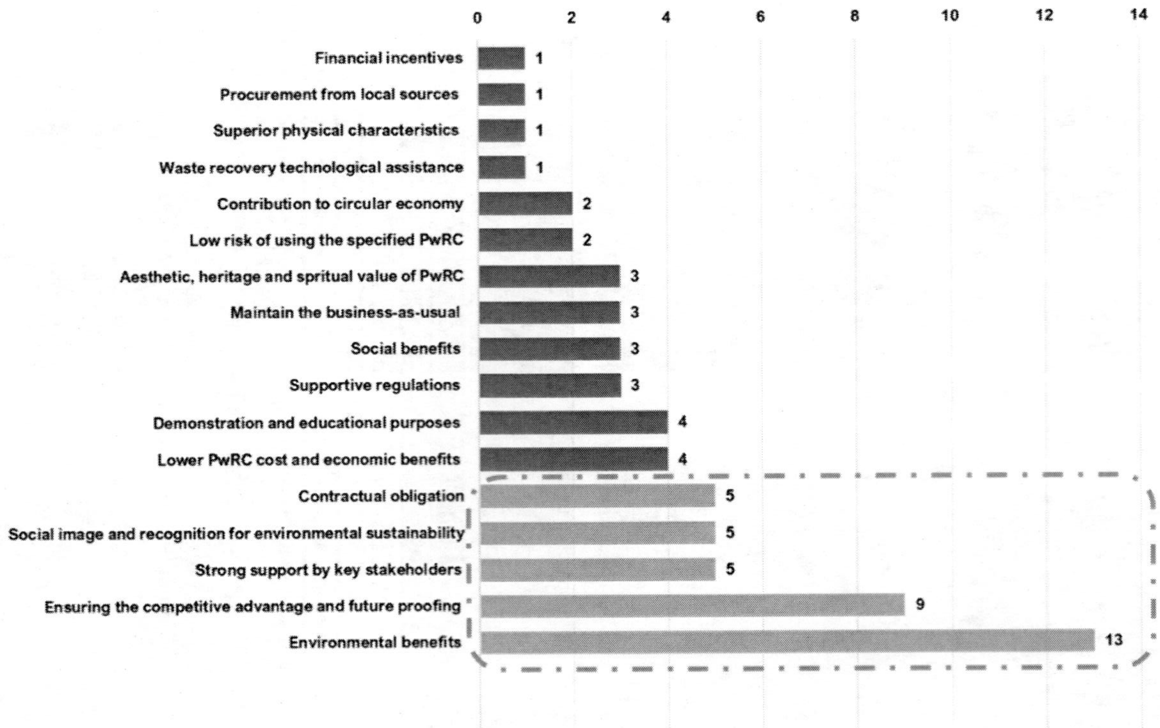

Figure 34. Frequency distribution of main motivations categories.

The motivational factors were categorised into 17 groups (Figure 34). The descriptive analysis showed that 'environmental benefits' (n = 13), 'Ensuring the competitive advantage & future proofing' (n = 9), 'strong support by key stakeholders' (n = 5), 'social image & recognition for environmental sustainability' (n = 5) and 'contractual obligation' (n = 5) were the top five categories of motivations for using PwRC in the four case studies.

The category of 'environmental benefits,' which represents about 17% of the total motivation factors, included reduced carbon emissions, less waste landfilling, project vision to be green, solving the issue of PwRC stockpiling and reducing virgin material extraction. Applying PwRC in the case studies resulted in the reduction of raw material extraction, waste disposal and some carbon savings. With respect to carbon emissions, however, as it was reported by C_2P_1:

> 'Sometimes you get environmental benefits in one area, but you're harming the environment in another area. So, there were trade-offs that sometimes did not work in favour of recycled materials.'

It is a fact that PwRC production and consumption require energy in transportation, processing and the application itself. Hence, construction environmental assessments need to consider the implications of PwRC utilisation to ensure it will deliver the environmental benefits as expected, without unintended consequences.

The second most important motivation category refers to the concept of future-proofing. As it currently stands, the Australian building and construction sector is preparing to shift towards adopting CE principles. In the near future, construction projects including public infrastructure projects will require contractors who can effectively handle PwRC. As C_3P_3 indicated,

> '...you also have to realise that when you win a contract as an alliance, a lot of the things that motivate the client to pick a certain team is based on their team, their people, and their attitude towards things like sustainability.'

Variation in participant views regarding the cost of PwRC was an intriguing finding of this research. In seven instances, the relevant expenses for using PwRC were viewed as an impediment, but in four instances, the participants were more interested in the associated economic benefits. This observation emphasises the unpredictability and multiplicity of factors that affect the final cost of the PwRC application. These variables include the project type, distance, application areas and financial incentives eligible to project owners that utilise PwRC. Similar results were achieved in the case of the role of regulations, which were found to be both supportive (an enabler) and unsupportive (a barrier).

7.4. Integrated Analysis of Barriers towards PwRC Application

The research participants reported 69 factors that influence the barriers to the broad application of PwRC in the selected case studies (Figure 35). The two stakeholder groups that provided the largest number of barrier factors in their responses to this question were the designers and builders.

To better analyse these factors, they were organised into 11 categories. In three instances, one response given by participants fell into two categories. The application of descriptive analysis to these categories showed that 'Unsupportive regulations' (n = 13), 'Inconsistency of PwRC quality, performance and warranty' (n = 9), 'Limited availability of PwRC' (n = 9) and 'Lack of expertise and understanding of PwRC utilisation' (n = 9) were the main categories of barriers towards using PwRC in the selected case study projects (Figure 36).

The barrier category referred to most, 'Unsupportive regulations,' represents several issues within current state waste regulations. The primary barrier is the complex and hard-to-achieve requirements stipulated in PwRC specifications. According to the interviews, these requirements practically limit the application of these resources in the sector. For instance, C_2P_2 indicated that:

> '...it is traditional specifications don't allow it...the pipe for instance. Well, the specification said it had to be a concrete pipe. So, if I was just to follow the specification that I can't do all the recycled asphalt, well, there was no testing regime there was not an accepted product. So, the challenge is to get the specifications to be invited innovation specifications traditionally.' [C_2P_2]

Previous studies also identified materials specifications as a major hurdle to not only the application of PwRC [21, 45], but also effective C&D waste management [74, 75], the creation and stimulation of markets for PwRC [76] and—more importantly—the implementation of CE principles [77]. In China,

Liu et al. (2020) [35] reported that waste regulations are the most influential factor in the C&D waste recycling industry that also affects other important factors. Meanwhile, it has been recognised that in Australia, 'overregulation, tough acceptance criteria, as specified in PwRC specifications, and increase testing' is the top barrier to effective C&D waste management [75].

The other complexity of current waste regulation in Australia pertains to the fact that regulatory frameworks are established on a state basis. As a result, each state has its own regulations and guidelines [78] and a range of materials specifications and standards are applied across Australian states and territories. Other barriers classified under this category were the frequency and routines of sampling of PwRC sourced from different C&D sites (Case Study 1), lack of obligation to use PwRC in infrastructure projects (Case Study 3) and difficulty obtaining permission to set up the temporary waste recovery plant and approval from city councils to use PwRC (Case Study 4).

Despite the increased C&D waste recovery across Australia [9], the issue of PwRC availability was frequently brought up in the interviews. Indeed, the procurement of quality PwRC that can meet applicable materials specifications and demonstrate consistent quality persists to be a major hurdle. This particularly was a pivotal issue in large-scale projects such as Case Study 1 where certain undersupplied PwRC were specified to be reused in the project in large quantities. Furthermore, the supply of PwRC was found to be difficult in areas located at a distance from the metropolitan areas. This issue was frequently noted by respondents from the builder category who were responsible for sourcing these materials for the study projects (Figure 35). C_3P_1 maintained that:

> '...it's another commercial challenge, which is around getting enough product at the right time at the right cost. So, once you've proven that it's okay to be used, then it'll be can we get it when we need it and get it for a reasonable price and it's competitive.' [C_3P_1]

Working with PwRC necessitates special expertise across the supply chain. Until this expertise and maturity become fully developed, the utilisation of PwRC is unlikely to become a business-as-usual practice in the sector. This barrier category represents issues such as the design team's poor experience and risk aversion, limited understanding and previous experience of applying these resources in construction and difficulty finding qualified contractors with the capability to use PwRC. C_1P_3 highlighted that

> 'I guess there's a process involved in finding the collaborators. That was not a problem, but it was something that took time and was part of the challenges.'

The last important barrier according to the responses refers to the quality of PwRC and the warranty thereof. This barrier represents a long-lasting issue in the sector that has been reported to significantly limits PwRC application and is linked with the risk of material failure, cost of the project, delays in project delivery and unwillingness to use them in the case projects. The PwRC quality issues emerge from various factors such as the inability to meet materials specifications requirements, carbon emission goals or expected performance, the existence of contaminants, limitations concerning sourcing materials with similar physical and visual quality, and requirements for special maintenance practices and warranties (Figure 35). This category was among the main concerns expressed by the client and supplier respondents. For instance, C_1P_1 contended that:

> '...if you have got performance requirements for, something that's structural or load bearing, or needs to be able to be cleaned in a certain way, or slip resistance, or whatever it might be, often, the easiest way to achieve those is with brand new products that have been, you know, factory tested, whereas you can't always guarantee that we've recycled or salvaged materials in particular.' [C_1P_1]

7.5. Integrated Analysis of Strategies for PwRC Application

During these interviews, 45 factors were identified as the main strategies to reduce the impact of challenges faced in these projects when using PwRC (Figure 37). The participants from the four stakeholder groups almost equally contributed to the identification of strategies to enhance using PwRC in construction projects.

As illustrated in Figure 38, these strategies were categorised into 13 groups, of which the top three categories are: 'effective education, investigation, and demonstration activities,' 'revising regulations in favour of utilisation of PwRC' and 'effective project management planning.'

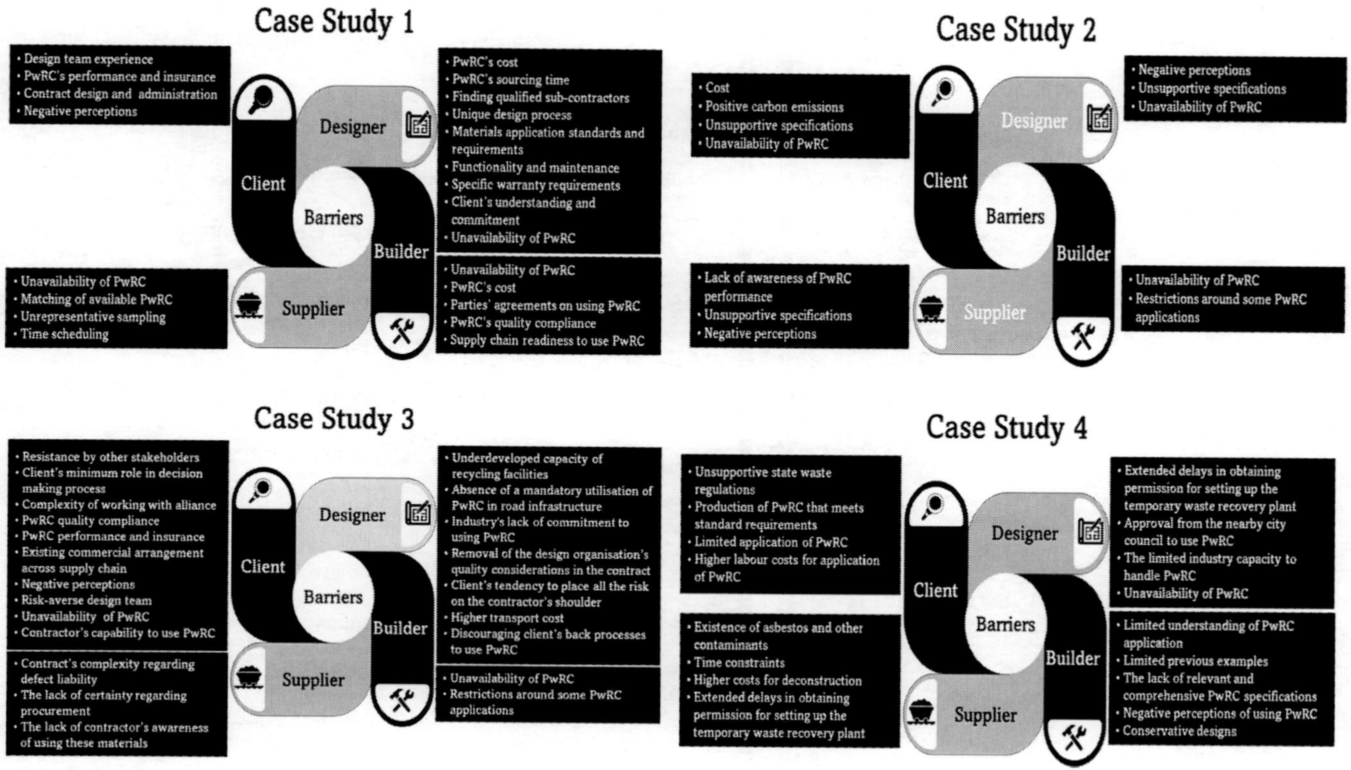

Figure 35. Barriers to using PwRC in the selected case studies.

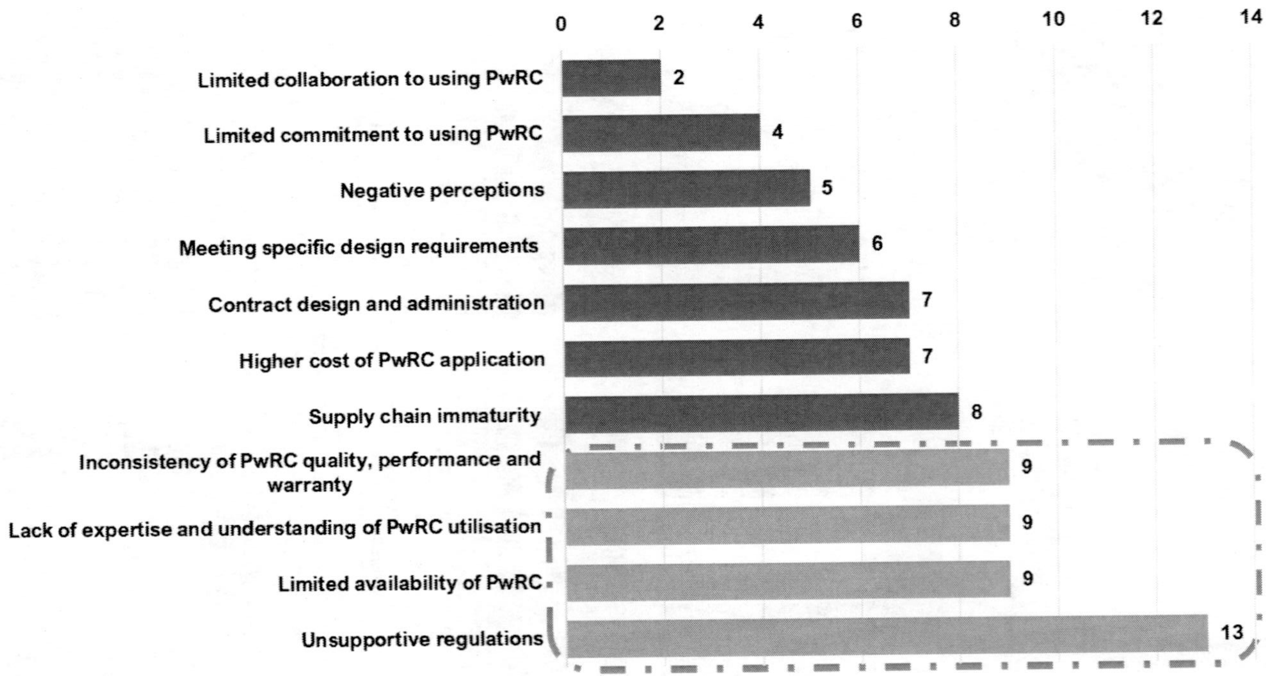

Figure 36. Frequency distribution of barrier categories.

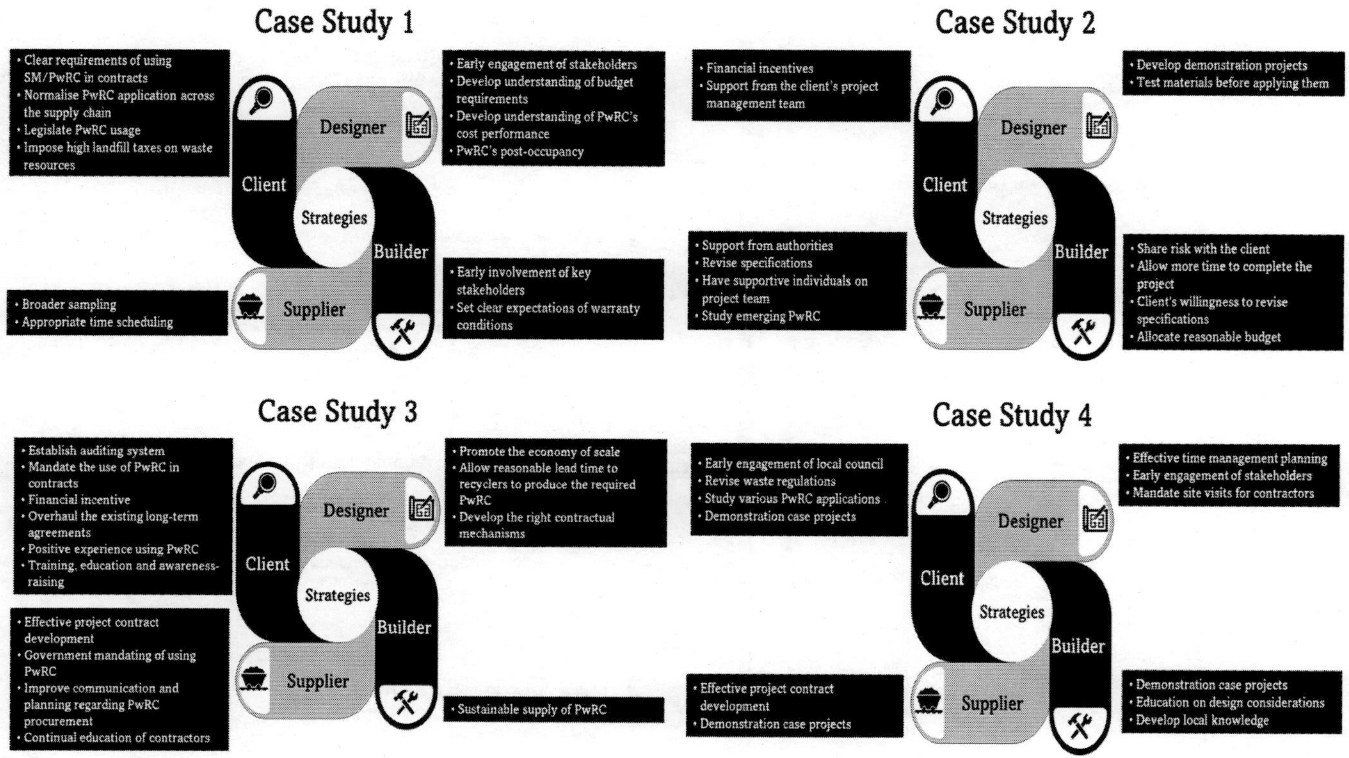

Figure 37. Main strategies to address barriers regarding the use of PwRC in construction projects.

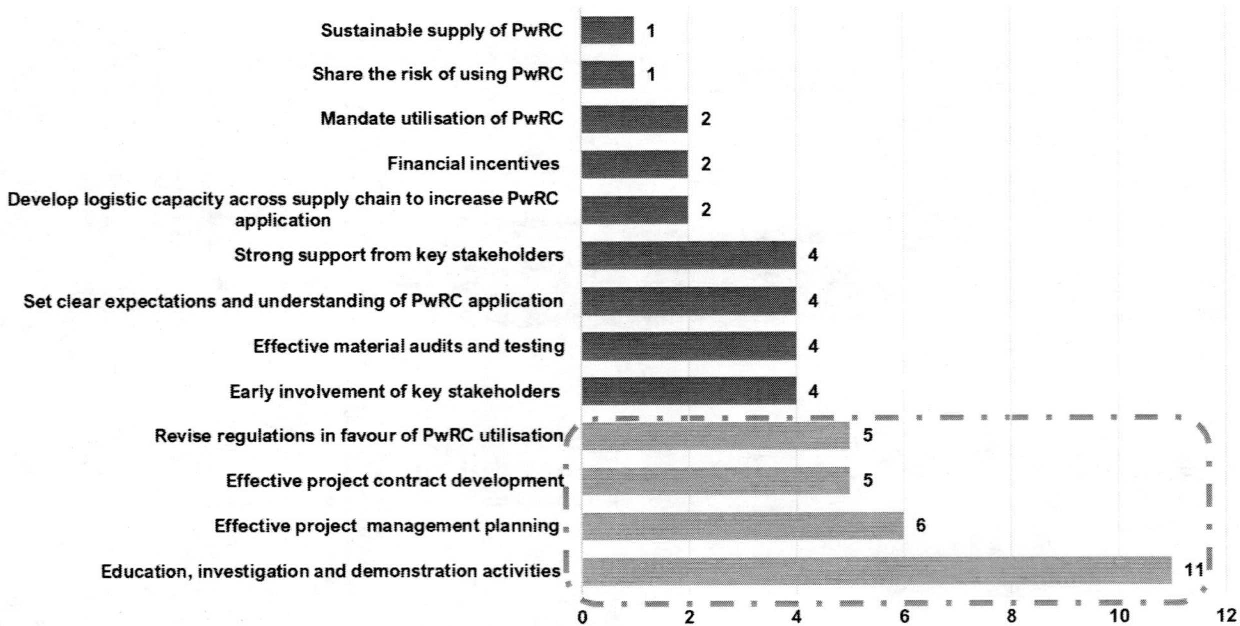

Figure 38. Main strategies to reduce the impact of challenges reported in the interviews

Efforts aiming to educate stakeholders together with encouragement and enforcement are often referred to as the three pillars of sustainable C&D waste management [79], and these motivate further uptake of PwRC [22]. This category represents several strategies: developing demonstration projects, studying emerging PwRC and new applications for these resources, continual education of contractors such as preparing mandatory site visits for contractors, developing local knowledge, and education on specific PwRC design requirements and complexities. In Australia, several educational opportunities target various stakeholders to improve their knowledge of C&D waste management including using PwRC in construction projects [80]. Demonstration projects can particularly drive the broad application of PwRC through attitudinal and behavioural changes among the key stakeholders [81, 82]. To put this into context, the results of a survey suggest that two in five respondents residing in Western Australia were not able to identify any projects where PwRC were used [20]. Furthermore, Liu et al. (2013) [83] found that the creation of demonstration projects for renewable energy applications in China has activated the entire market—i.e., the demand and supply sides—and as a result, such applications have become more popular. This fact was clearly reflected by C_4P_1, who stated:

> '…the purpose of these projects for us is to be able to demonstrate to the industry. Whereas as state government, obviously, we want to be showing that we're pushing the boundaries and trying to, I suppose provide demonstration projects that show what can be done within a normal commercial environment.' [C_4P_1]

Knowing the importance of this strategy, the owners of all four case studies offered site visits to provide learning opportunities for individuals working in the relevant fields.

The second broadly referenced category deals with project management planning strategies such as effective time planning, communications and reasonable budgeting. C_4P_3 indicated that:

> 'It's ultimately about good planning, understanding the timeline implications. Certainly, that means the time frame challenge is you would hope that early engagement with regulators helps identify risk and ensure that all parties understand what the intention of the project is.' [C_4P_3]

This observation highlights the importance of the project management team to the successful completion of PwRC-using projects. Project managers are found to be among the top four out of 20 stakeholders recognised as having an impact on the successful implementation of C&D waste management in Australia [84].

7.6. Comparative Analysis of Participants' Responses on the Application of PwRC

Further analyses were conducted to compare the major categories created for reporting different determinants of PwRC application (i.e., motivations, barriers and mitigation strategies) across the case studies as well as the stakeholder groups. Further analyses were undertaken to examine and compare the categories reported in the four case studies and by various stakeholders. The analytical focus was placed on the categories with the largest frequency in each case study and stakeholder group. In some instances, stakeholders selected more than one category, hence expanding the number of categories compared.

As can be seen in Table 14, the results showed that participants in each case study prioritised the defined categories differently. Such findings reveal internal and external factors can impact construction projects in different degrees or directions. It can be concluded that while the findings based on the aggregated data provide useful insights into PwRC application, precise planning for using these resources needs to be done individually and subject to contextual variables, including evident and latent constraints and opportunities.

A similar trend was evident in the comparison of categories between the stakeholders (Table 14). The findings show that stakeholders did not have similar perceptions with respect to the PwRC application. In terms of barriers, only two out of five categories were selected by all stakeholders. This ratio was one to five and one to six for motivations and strategies, respectively. For instance, 'environmental benefits' was the most important factor only for clients and designers. This finding corroborates previous research in which researchers observed variations in the perceptions of various stakeholders about enablers of and barriers to using PwRC in construction projects in Japan [44] and New Zealand [85]. Furthermore, the research filled the gap that was previously identified by Udawatta et al. (2018) [86] concerning the need to study the variations involved in conducting waste management practices due to different project types and among different stakeholder groups in Australia.

Conclusion

In Chapters 3-7, factors influencing the use of PwRC were investigated through a multiple case study approach in the Australian context. As per the integrated analysis in this chapter, the top four barriers to using PwRC in construction projects are 'unsupportive regulations,' 'limited availability of PwRC,' 'lack of expertise and understanding of PwRC utilisation' and 'inconsistency of PwRC quality and performance and issues around the warranty.'

The research findings will have policy implications as the 'unsupportive regulations' were recognised as the top barrier. The current climate of the waste recovery system in Australia, and possibly all around the world, means governments are being urged to devise policies that facilitate the utilisation of PwRC and drive a CE in the building and construction sector. As such, policymakers can use the information provided in this study to develop policies that provide a level playing field for all parties operating in PwRC supply chains. For instance, informed policies can incentivise the production and purchasing of PwRC in the sector.

According to the responses from the research participants, the barriers identified in this study are to be primarily addressed through 'education, investigation, and demonstration activities' and 'effective project management planning.' Further research should investigate how the outlined strategies may change stakeholders' attitudes and result in the optimised use of PwRC in construction. In addition, the policy reform initiatives must pay particular attention to these measures as part of a road map leading to the optimised use of PwRC.

Additionally, the two main motivations for using these resources were found to be 'environmental benefits' and 'ensuring competitive advantage and futureproofing.' Accordingly, new policies can give the motivating factors found in this study a greater weight in the equation of PwRC utilisation. The research also has a practical contribution to the industry through a suite of actionable strategies to increase the uptake of PwRC. The two major motivations, referring to two of the three pillars of sustainability (economy and environment), provide a basis for organisational changes to ensure the achievement of sustainability in construction activities.

Table 14. Summary of comparative analysis of the PwRC application determinants across case studies and stakeholders

Case study	Barriers	Motivations	Strategies
1	• Inconsistency of PwRC quality, performance & warranty	• Ensure the competitive advantage & future proofing • Social image & recognition for environmental sustainability • Strong support by key stakeholders • Supportive regulations	• Set clear expectations & understanding of the PwRC application
2	• Unsupportive regulations	• Ensure the competitive advantage & future proofing • Environmental benefits	• Strong support from key stakeholders
3	• Contract design & administration	• Ensure the competitive advantage & future proofing • Social image & recognition for environmental sustainability • Strong support by key stakeholders • Supportive regulations	• Effective project management planning • Effective project contract development
4	• Unsupportive regulations	• Environmental benefits	• Education, investigation & demonstration activities
Stakeholders			
Client	• Inconsistency of PwRC quality, performance & warranty	• Environmental benefits • Supportive regulations	• Education, investigation & demonstration activities
Builder	• Limited availability of PwRC • Unsupportive regulations	• Contractual obligation	• Education, investigation & demonstration activities

Case study	Barriers	Motivations	Strategies
Designer	• Unsupportive regulations	• Ensure the competitive advantage & future proofing • Environmental benefits	• Early involvement of key stakeholders • Education, investigation & demonstration activities • Effective material audits & testing • Effective project management planning • Set clear expectations & understanding of the PwRC application
Supplier	• Contract design & administration • Inconsistency of PwRC quality, performance & warranty • Supply chain immaturity • Unsupportive regulations	• Ensure the competitive advantage & future proofing • Maintain the business-as-usual	• Education, investigation & demonstration activities

The thematic and descriptive analyses revealed that the study factors might affect the utilisation of PwRC to a different degree and/or in different directions. Hence, while the aggregated data in this study provide useful insight into PwRC application, precise planning for using these resources is subject to a wide array of contextual variables, including evident and latent constraints and opportunities.

Chapter 8

The Role of Certification Schemes in the Adoption of PwRC

8.1. Introduction

Although it has been determined that the utilisation of PwRC is a sustainable solution for the management of C&D waste, there are still issues regarding the quality, contamination, and performance of these products.

Among the many potential solutions that have been proposed to address these worries, product certification stands out as one that could actually be implemented. It is possible for there to be a rise in the utilisation of PwRC in the building sector if there were to be a product certification that was granted after material testing as well as quality control.

As different supply sources will cause variation in quality, stricter quality controls for PwRC are required [87]. According to industry and government reports, it makes economic sense to apply current certified recycled C&D waste materials in large infrastructure projects. In a research report [88] commissioned by the Waste Authority, C&D recyclers reported that their certified road-base products are valued at $5-6 per tonne, while virgin quarried materials are valued at up to $20 per tonne. Furthermore, the report indicated the figures for certified clean-fill products and their virgin alternative are $3 and $10 per tonne. These numbers indicate that the existence of such programmes could help create confidence among stakeholders, culminating in a market shift in favour of recycled-content products.

Nevertheless, the implementation of certification systems for recycled products comes with a few obstacles to overcome. Before anticipating widespread adoption of this solution among the various stakeholders in the public and private sectors, these challenges need to be carefully studied and addressed appropriately.

This initiative has been implemented worldwide, in Europe, the UK, the USA, and South Korea. These schemes reassure the end-users of satisfactory utilisation of recycled products and include parameters such as energy use, recycled content, and air and water emissions from manufacturing, disposal, and use. The initiative has also been recently introduced to Australia, but current programs are still at the early stage of development and adoption.

This chapter is organised into five sections, each of which contains information about the RPC schemes. The next section covers the key functions of the certification schemes for PwRC and their primary challenges and enablers of its widespread applications are examined. In the following sections, their applications in the Australian building and construction sector are explored, followed by a discussion of a framework designed to facilitate its deployment in the building and construction sector. The conclusion section presents the limitations of research, further studies and the way forward.

8.2. Review of Recycled Product Certification Schemes

8.2.1. What is Recycled Product Certification (RPC)

Recycled product certification provides verified information on recycled products' quality, performance, environmental friendliness, and safety. An RPC that is awarded following material testing, plus quality control, can increase the adoption of recycled materials.

All businesses operating in a supply chain may benefit from developing RPC to show the specifications of their products required by buyers [89] (Figure 39). For instance, for builders, RPC is an efficient management of information reducing the workload in a tender process. Utilising RPC will allow builders to manage any risk that may arise from the usage of PwRC. In this circumstance, clients and contractors will bear the risk equally. Such programs also enable sustainable procurement policies in private and public sectors domestically and develop overseas markets for states that recognise certain certificates. For clients, it offers options for those who are willing to work towards sustainable development [90]. For suppliers and clients, RPC schemes will serve as a marketing tool to demonstrate the sustainability of their products and projects, respectively.

It is highlighted that the classification of recycled products and aggregates into easily comprehensible categories, together with correct certification, facilitates future client purchases [91], as they will be able to acquire a product suited to its intended use (e.g., structural concrete may use recycled aggregate of class A, whereas recycled aggregate of class D may be used for subgrade, in road construction).

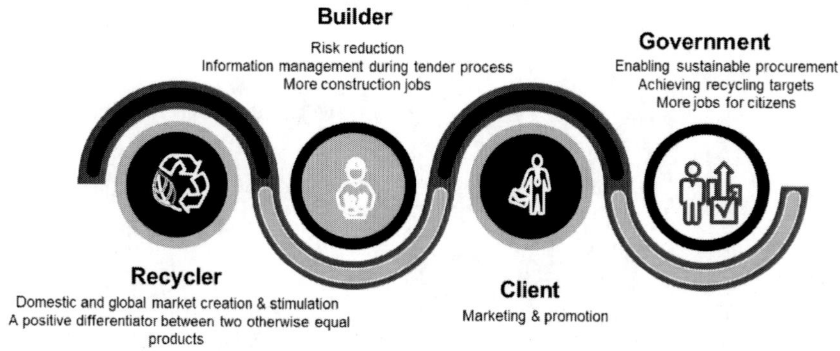

Figure 39. The advantages of RPC for recycled products supply chain participants.

Currently, there are two categories of marketable recycled products: non-certified and certified. Non-certified recycled products constitute the majority of the output of recycling plants. However, certification is of the utmost importance due to consumers' increasingly stringent needs for recycled products of a specified and assured quality.

It is important to understand that product standards and eco-labels are distinct. The purpose of eco-labels is to inform consumers about the environmental and human health impacts and potentially ethical labour practices associated with the products they purchase. The majority of pass/fail type eco-labels do not give consumers real information about the product, but rather suggest that it is among the top 10 to 20 per cent of its product category from an environmental standpoint [90].

8.2.2. Global Application of Certification for Recycled Products

There is growing empirical evidence that proves the effectiveness of RPC in driving the market for PwRC. In the EU, it is unlawful to use PwRC that have not been certified [92]. A recent study conducted in China [19] revealed that despite the low frequency of RPC use in the 52 studied cities, it had the greatest impact on the market for recyclables compared to other policies. In light of the study's findings, the authors urge that, among other policy instruments, RPC should be given the highest priority. Participants in an Australian study [20] on the usage of PwRC in the landscaping sector ranked product certification as one of the four policy instruments most likely to enhance the adoption of PwRC in Western Australia.

Figure 40. Examples of RPC programs in different countries.

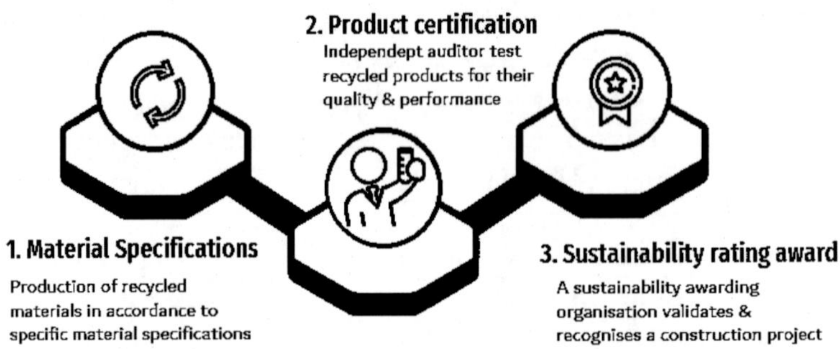

Figure 41. The process of sustainability assessment for rating a construction project.

In a few countries, RPC is progressively implemented under several names. In China the main ones are Green Product Label [93] and Green Building Certification [93], in Luxemburg, it is 'Bridges and Roads Materials Testing Laboratory Certification [94], in South Korea, it is Korea Eco-Label (G-SEED) [95], and in the UK it is Building Research Establishment (BRE) Certificate, in Belgium, it is CEN KEYMARK. The following figure presents a snapshot of various RPC versions in different countries (Figure 40).

In the majority of states that use these certifications, there are three processes involved in awarding a project (rather than a product) a sustainability grade (Figure 41). In the first step, materials are created in accordance with the specifications normally issued by local authorities for a variety of applications. In the second step, independent auditors and organisations evaluate and grade the PwRC based on their quality, performance, and level of purity. In the last step, the sustainability rating awarding organisation grants credits in various forms, such as points accumulating towards a given number of stars, to a construction project in recognition of the project owners' efforts towards the usage of PwRC and overall sustainability.

8.2.3. Primary Drawbacks of RPC

Despite the lengthy history of product certifications [90], their full implementation in the building and construction sector is rather recent. Consequently, it is important to note that the number of publications detailing the problems associated with their actual use is minimal. The review of pertinent literature found that the implementation of product certification procedures for recycled C&D waste materials is hindered by several obstacles. Figure 42 depicts that these include the process being costly, the costs being currently unjustified, self-certification, the lack of alignment between RPC and sustainability rating systems, operation of unreliable certifiers and lack of collaboration between parties involved in RPC application.

A report by Equilibrium [96] indicated that in Australia products with recycled content are not independently validated through robust support testing. Product certification is cost-intensive, and with little up-take and a lack of opportunities to use these products, justification of the expense of certifying these products is difficult. Gaffar (2019) [92] indicated that mandatory product certification expenses in the EU add to the price of the material and may negate any cost savings from reusing it.

Self-certification is also utilised to establish the quality of PwRC. Recyclers and providers of these materials often perform self-certification. Although it is less expensive, the certification does not assure that end users may safely utilise PwRC.

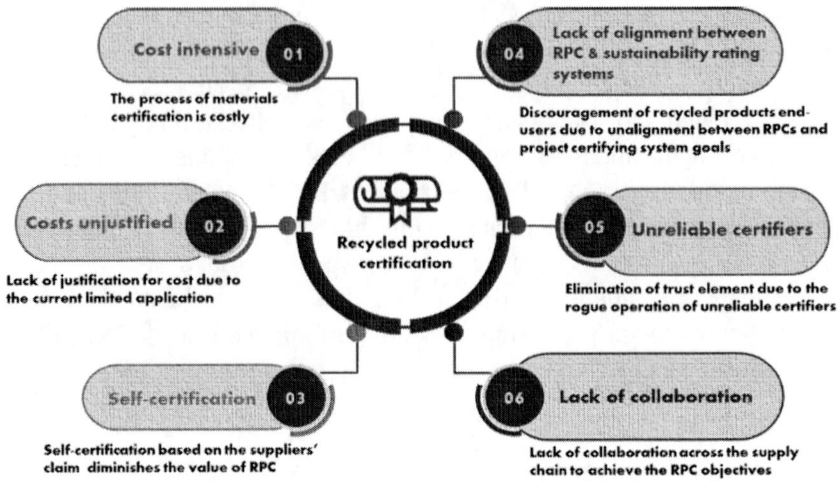

Figure 42. Main cons of the application of recycled product certification.

8.2.4. Primary Enablers of RPC

Six categories of RPC enablers are identified following a comprehensive literature review. These classifications are illustrated in Figure 43. Effective quality monitoring and certification of recycled products by suppliers are required to instil and maintain stakeholders' trust in the PwRC application. However, this must be accompanied by more vigorous government's action in the form of legislation and standardisation [91]. The government should also initiate promotional activities via the media to change the general public's attitudes towards recycled products. The government may also organise some regular training on changing the industry stakeholders' attitudes [97].

Furthermore, contractual obligation or otherwise client requirement to use certified recycled products is another enabler for the implementation of RPC [91]. Currently, in the EU, it is illegal to use recycled content in construction materials that are not certified [92], which has resulted in more frequent uptake of these programs in construction projects across the European territory. According to a report by WSP [90], the advantages of mandatory RPC include the eradication of rogue traders, greater confidence in the data provided for all parties, less room for manipulation, and the credibility required to make the RPC valuable.

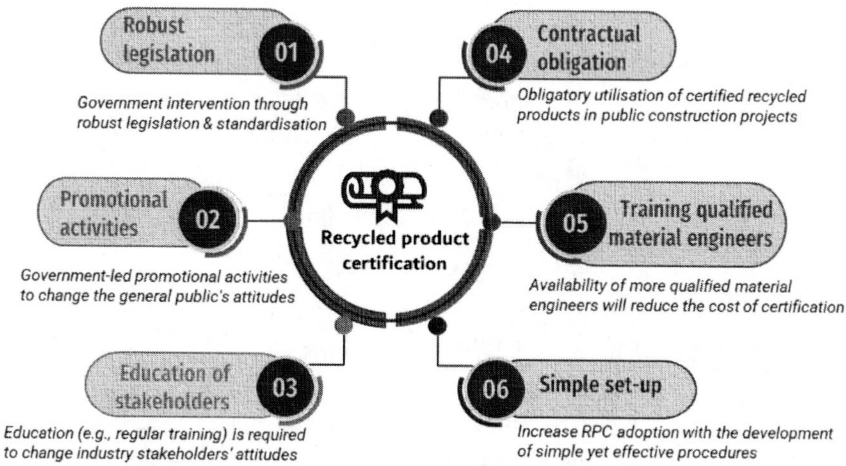

Figure 43. Main enables of the application of product certification for recycled products.

The fifth category of enablers refers to the availability of qualified material engineers at reasonable prices. According to anecdotal evidence, these professionals are scarce and expensive to employ. By educating more individuals to become trained materials inspectors, not only will jobs be created, but audit costs will also be reduced. The final category addresses the simplicity of certification methods for recycled products. The waste recovery industry's interest in RPC will be piqued by certification processes that are simple yet effective and require a fair amount of administrative work [90].

8.2.5. Australian Regulatory Framework and Recycled Product Certification

To comprehend the Australian government's perspective on RPC, national, state, and territory legislation and statutes were exhaustively analysed. The results indicate that there is no mention of certification requirements for PwRC to be utilised in construction projects in any Australian regulations.

Recycled product certification is not also recommended among many strategies stipulated in the 2018 National Waste Policy [98]. In addition to the current primary and secondary legislative documents that govern different waste streams, Australian jurisdictions have also produced waste strategy documents. These documents are guiding efforts and legislation in each jurisdiction towards better management of waste including C&D waste. Waste strategy documents are an important part of waste management governance in Australia [78]. The examination of these documents revealed that none of the states or territories that currently have a waste strategy document (including New South Wales, the Australian Capital Territory, Queensland, South Australia, and Western Australia), support or introduce the concept of recycled product certification program.

8.3. Analysis of Existing RPC Programs

In the absence of precise information about the operation and growth of several entities issuing certificates for PwRC in the building sector, the following literature review provides a non-comprehensive overview of product certification in Australia.

An Environmental Product Declaration (EPD) [99] is a standardised way of quantifying and reporting the potential environmental impacts of a product or system based on ISO 14040 and ISO 14044 compliant life cycle analysis (LCA). EPDs are produced according to a consistent set of rules – Product Category Rules (PCR) – that define the requirements within a given product category. These rules are a key part of ISO 14025 (ISO 2006) as they enable transparency and comparability between EPDs. The internationally acceptable standard and common PCR for construction products is EN 15804.

Good Environmental Choice Australia (GECA) Standard [100]: This pass/fail standard identifies environmental, quality, regulatory and social criteria that the top products sold in the Australian marketplace can meet in order to be recognised by GECA as 'environmentally preferable.' This standard identifies environmental, quality, regulatory and social criteria that the top products sold in the Australian marketplace can meet in order to be recognised by GECA as 'environmentally preferable.' The primary purpose of this standard is to define environmental performance criteria for products made from recycled content materials (Table 15). Recycled products may be environmentally preferable based on reduced demand for virgin materials and reduced amounts of waste going to landfill. The standard is applicable to the following categories of recycled/reused content products: glass, plastics, rubber, timber, fill and foundation materials such as sand and gravel.

ViroDecs™ [101] is developed by Holcim as an EPD that covers its normal classes of concrete ranges nationwide. Launched in 2019, the ViroDecs™ EPD enables designers and builders to make more informed choices about materials based on LCA of their environmental impacts. Completely transparent and third-party verified, the ViroDecs™ EPD provides a range of environmental indicators for Holcim's ready-mix concrete products. Since the launch of the ViroDecs™ EPD, Holcim has been recognised by the Green Building Council of Australia for its industry leadership which continues to shape the way the construction sector designs and builds.

A minimum of 50 per cent of the product by weight must be made up of one of, or a combination of, these materials. This, or these, material/s must be used to show compliance to Criterion 3. Other products that do not fit the above categories may be considered for certification provided the product satisfies all relevant requirements. Other recycled content products may be added to the scope at a later date.

Global GreenTag [64] is one of the certifying operators that include RPC in the Australian and global context. The organisation independently assures that every product is fitness tested and certified under one of several leading certification programs that use standard scientific methods and as a prerequisite, requires physical materials testing under relevant standards by qualified ISO 17025 laboratories. Each of the certifications below represents a progressive increase in the assessment detail, complexity and holistic nature of the certification outcomes and cost:

Global GreenTag's Manufacturer Claim Verification (McV) is the simplest form of RPC. It undertakes supply chain and ingredient audits to provide basic certification that warrants a manufacturer claim is valid based on current contracts, and accounting documents such as invoices, weighbridge dockets, and bills of lading together with testing. It results in a simple 'Tag' with the claim reproduced, e.g., 'Minimum 80% Post Consumer Recycled Content.'

GreenTag's *GreenRate*™ certification is a Green Building Council of Australia, New Zealand and South Africa and Infrastructure Sustainability Council recognised product rating system designed to meet the requirements of the Green Star® 'Sustainable Products' and 'Responsible Products Framework' and 'IS' rating tool materials credits respectively in the various versions of those rating tools. It recognised levels of compliance to A, B and C performance levels each recognised distinctly within Green Star®.

Based on Full Disclosure of ingredients to 100ppm, it undertakes a 3-stage risk assessment process considering Health and Ecotoxicity Hazards and other Priority Assessment Criteria. It looks at different aspects of the product and its ingredients at each key life-stage stage (ingredient, worker and in-use) and for various issues including; aspects of resource use (post-consumer recycled and rapidly renewable content), waste reduction, availability of greenhouse and water footprints.

LCARate™ is a comparative rating system that uses full whole-of-life (cradle to fate) LCA to compare the environmental, health and social impacts and benefits of the product by comparison to a commonly available 'worst case business as usual (BaU)' product that performs the same function. The rated outcomes result in Bronze, Silver, Gold or Platinum levels of recognition and these are achieved by awarding of point scores between -1.00 to +1.00 (effectively a score out of 200) across six sustainability assessment criteria; Health & Ecotoxicity, Building Synergy, Biodiversity, LCA, Climate Impact and Corporate Social Responsibility including ethical labour conditions.

Also based on Full Disclosure of ingredients to 100ppm, it considers products and their ingredients in greater scope and detail using full ISO 14040 and 14044 compliant LCA as well as the same Health and Ecotoxicity risk assessment process as in GreenRate™. RPC is undertaken against a worst case sourced BaU virgin material with discrete, material specific environmental and human health risk assessment and where appropriate on-site audit, together with all aspects of resource use (post-consumer recycled, rapidly renewable content), waste and water reduction and product specific ISO 14067 compliant carbon together with an embodied water assessment.

GreenTag's *NaturePositive⁺ Declaration* is an emerging Certification that takes account of the issues that are dealt with within LCARate but then adds a Circularity Indicator that includes a material quality score and a unique Life Cycle Benefit Assessment (LCBA) combined with a product supply chain modern slavery report and rating.

This takes RPC to a new level by not only providing certified ratings but by presenting a full transparency reporting declaration that is third-party audited. It then provides a framework to show how biodiversity and carbon co-benefit offsets can be engaged to deliver nature-positive outcomes with GBCA and National Biodiversity Certificate compliant outcomes (subject to finalisation of the GBCA Code and National legislation respectively, yet to occur).

8.4. Application of RPC: Analysis of Four Case Construction Projects

To understand the industry key stakeholders' perceptions of the usability of RPC, the individuals, who were recruited from four recently completed construction projects, were interviewed. The interviewees were from four key stakeholders (i.e., client, builder, designer, and supplier) as described before. The relevant questions in this section aimed to explore four aspects of PRC application in the building and construction sector. These include the interviewee's awareness of this initiative, its actual application in the case project, whether they support the application of RPC, and their descriptive opinions on its usability.

The following table provides a snapshot of the responses to the questions aiming to explore the specific aspects of the RPC application in the case project as well as in general. Among the four case studies, only one project (i.e., Tonkin Gap and Associated Works) employed some sort of RPC to endorse the utilisation of PwRC. The process of certification was done internally by the client organisations. This certification was developed under the Roads to Reuse (RtR) initiative and is currently used in several WA government projects to ensure the quality compliance and safety of PwRC applications in the state road infrastructures. In Case Study 2, the client organisation conducted several tests (e.g., UV stability to fire rating to durability impact testing) to ensure that the PwRC used in the project would have the quality that is acceptable to the Victorian Department of Transport.

In terms of the stakeholders' awareness, the findings suggest that 9/16 interviewees were aware of the existence of such certifications for PwRC (highlighted in green). Whereas 7/16 interviewees indicated they had no or limited prior knowledge about their application in the industry (highlighted in red/grey). A similar trend was also observed in the way that the respondents provided their responses with respect to their support of their application to the construction projects in Australia. Meaning that only 9/16 interviewees had positive attitudes towards their application in the industry.

Table 15. Criteria used by GECA to certify recycled products

No	Description	Demonstration of Conformance
1	The scope of this standard is applicable to products that are manufactured partially or fully from recycled content products and sold on the Australian market.	A brief description of the product(s) or product range as they apply to the scope of this standard.
2	The product meets or exceeds the requirements of the relevant Australian or International Standard for its intended application; or, where there is no applicable quality standard, the product must demonstrate fitness for purpose or market acceptance or suitability or quality.	Independent audit or test reports, or report from an independent organisation that demonstrates fitness for purpose, market acceptance, suitability or quality.
3	The recycled content component/s3 must comply with at least one of the below recycled content requirements. Glass (100%), plastics (90%), rubber (90%), timber (100%), fill and foundation materials such as sand and gravel (100%)	Declaration of conformance supported by relevant quality control and production documentation.
4	Adhesives used in the manufacturing of the product shall not contain alkylphenol ethoxylates or other alkylphenol derivatives, phthalates (that at the time of application are classified with risk phrases H360, H361), or halogenated volatile organic compounds.	Safety data sheets (SDS) of the adhesive or A copy of relevant GECA or other relevant ecolabel certificates.
5	Treatment of wood products Wood, including engineered wood products and natural materials, must not be treated or impregnated with fungicides and insecticides that are listed by International Agency for Research on Cancer (IARC) 2B.	List of all agents used (including chemical abstract services (CAS) numbers and names) to treat wood and other natural materials (to enable spot checks). DoC 12.2: Signed declaration from an Executive Office stating that any pesticides used are not classified by the IARC as 2B.

Table 16. The cross-comparison of findings in the four case studies

		Client (P1)	Head-contractor (P2)	Designer (P3)	Supplier (P4)
Application	C. S 1	No application			
	C. S 2	No application			
	C. S 3	Internal testing of PwRC through roads for reuse			
	C. S 4	No application			
Awareness	C1	**Yes**	Limited knowledge	No	**Yes**
	C2	Limited knowledge	No	**Yes**	**Yes**
	C3	**Yes**	**Yes**	**Yes**	**Yes**
	C4	No	No	**Yes**	Limited knowledge
Support	C1	**Yes**	Did not specify	Did not specify	No
	C2	**Yes**	**Yes**	**Yes**	No
	C3	**Yes**	**Yes**	No	**Yes**
	C4	**Yes**	No	**Yes**	Did not specify
Opinion	C1	Confusing/not clear	Did not specify	Not sure how and what is measured – complicated	Not possible due to variations in the origin of waste resources
	C2	Did not specify	Good way to reduce risk	Helpful to make sure it performs as it is intended	The supplier has its own PwRC certifications, the existing schemes can create some barriers
	C3	Useful for ensuring the PwRC, improving the circular economy and supporting the local waste recovery industry	Good for some PwRC and it will minimise the waste landfilling	Specifications are good enough, unnecessary extra costs and not sure if they are adding value	It is required to maintain high-quality standards which builds confidence in the product
	C4	To ensure PwRC do not pose a hazard to human health	It diverts their focus which is only placed on the design and the sustainability rating program they used	Is required to build confidence in PwRC for end-user but should not be costly	Did not specify

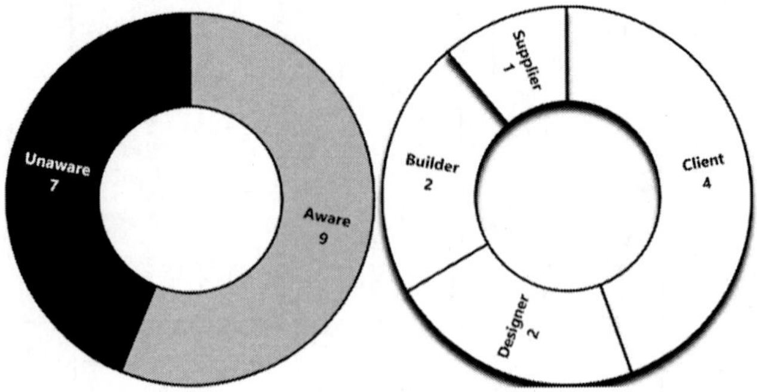

Figure 44. Awareness of RPC; left, overall (n/16) and right, by different stakeholder groups (n/4).

In terms of the stakeholders' awareness, the findings suggest that 9/16 interviewees were aware of the existence and objectives of RPCs. Whereas 7/16 of the interviewees indicated that they had no or limited prior knowledge about their application in the industry. As presented in Figure 44, suppliers and designers were the two stakeholder groups that had more knowledge of RPC application.

8.4.1. Stakeholders' Support for RPC Application

The response to whether or not research participants supported the application of RPC showed that a larger portion of them considered RPCs a useful tool during purchasing PwRC (Figure 45). Stakeholder analysis suggests that clients and suppliers were the most and the least interested stakeholders in using these schemes, respectively.

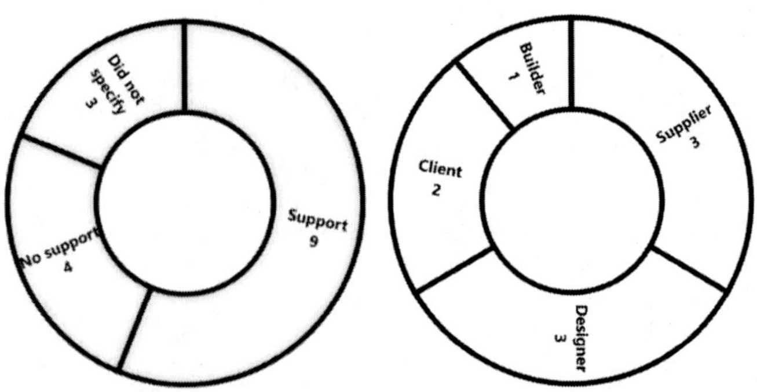

Figure 45. Support for RPC application; left, overall (n/16) and right, by different stakeholder groups (n/4).

For instance, C_2P_2, the builder stakeholder, said that RPCs are effective in reducing the risk of using PwRC in construction projects:

> "One of the impediments to using recycled products is that [whether or not] these things are going to perform…to some extent, we're taking on the risk that these things are going to perform. But if some certification body takes on that risk for us, and we can say well, we're relying on their certification, well that would be removing any barrier to using the product. That's what we need, we need to know that these things are going to perform".

The following sections provide an overview of the application of RPC in each case study to further detail the stakeholders' perceptions on using them in construction projects.

8.4.1.1. Case Study 1 – Brickworks Shopping Centre

In this case study RPC was not utilised and the project team focused on obtaining a sustainability tool so as to be recognised for sustainability including the use of PwRC. Particularly, the project used the Living Building Challenge [58] and GBCA Green Star to demonstrate the utilisation of these resources in Brickworks Shopping Centre. The former that has seven petals (themes) required the project to apply SM/PwRC in 30 application areas. In addition, based on the building's gross floor area, the project team had to use one recycled material per 500 m^2 in the project.

As part of this challenge, the suppliers of materials had to provide product certification that show their level of environmental sustainability. However, according to the responses, for SM and PwRC no RPC was used except for the materials manufacturers' statements. As per Table 16, two out of four individuals interviewed had limited to no knowledge of the existence of RPC programs. Furthermore, only one interviewee, the client [C_1P_1], explicitly supported the application of these programs to ensure the safe utilisation of PwRC. In terms of the limitations, C_1P_1 indicated two limitations which are related to the abundance of these programs and the lack of transparency about their requirements.

> 'I think one of the challenges is that there are obviously so many different kinds of certifications, a lot of people get quite confused as to what they will mean. I think the other challenge, is that some of the certifications are a little bit opaque. So, you often find certification, you just get some

sort of points or star rating system, but you don't really even know what that even means, you know, Level A or number three, four. It doesn't tell you anything it just tells you that there's some sort of rating system' [C_1P_1]

The other issue that was raised by the project's designer C_1P_2, was a minimum quantity of SM/PwRC to justify the cost of testing under any RPC. They believed that the use of PwRC might be limited to a certain percentage that might not justify undergoing the same level of testing as it would in the case of other materials produced in large quantities.

'But there's new carpet that has a component of recycled material, which would have all of those other codes, it would pass because they can actually produce it in quantities large enough to validate the cost that they have to, you know, to undertake all of that testing, which is like we're talking, you know, for one test could be $50,000' [$C_1P_2$].

The supplier of recycled timber in this project mentioned that their organisation is never able to get certification for this product due to its nature being a live product that moves over time.

'It is a natural product, especially with timber. also noting that there can be movement with timber, it's a live product, it does move so it moves to its surroundings and stuff like that.... in terms of certificates and certifications and stuff like that we could never warrant or give anything'[C_1P_4].

8.4.1.2. Case Study 2 – Mordialloc Freeway

The project owner of the second case study did not use any RPC. However, they get suppliers' self-certified documents for certain materials such as plastic noise walls. Furthermore, they conduct a wide range of testing on these resources to get approval according to the Victoria Department of Transport Specifications. The analysis of responses suggested that only two out of four interviews (i.e., C_3P_3 and C_3P_4) from this project were aware of the existence and application of RPC in construction projects. However, except for the representative of the supplying organisation, other interviewees expressed their support for using these certifications (Table 16). For instance, the second participant (C_2P_2) said that RPCs are effective in reducing the risk of using PwRC in construction projects.

> 'One of the impediments to using these recycled products is that [whether] these things are going to perform. ... to some extent we're taking on the risk that that these things are going to perform. But if some certification body takes on that risk for us, and we can say well, we're relying on their certification, well that would be removing any barrier to using the product. That's what we need, we need to know that these things are going to perform' [C_2P_2].

In terms of the RPC limitations, two interviewees (C_2P_2 and C_2P_4) responded. C_2P_2 identified the premium cost imposed on PwRC due to the application of these certificates as a significant barrier to their adoption. C_2P_4, however, questioned their usefulness by referring to the potential barriers they might create and their needlessness in the presence of materials specifications.

> 'I think the problem would be if you if the product starts coming at a huge premium because of the certification process, well, that might be problematic' [C_2P_2].
>
> 'I think a lot of the time, unfortunately, they can get a little bit elaborate and actually create more barriers and really, the value of some of them, in my opinion, is a little bit questionable. So I think we need to probably think out of the box a little bit. You know, we have specifications for a reason. We need to maybe think about how we make those specifications more responsive, more agile…I think that would be far more useful than creating certification on top of certification to say that a product is potentially viable, that has nothing to do with the actual road building specifications' [C_2P_4].

8.4.1.3. Case Study 3 – Tonkin Gap and Associated Works

In this case study, an RPC was employed to verify the quality and compliance of the PwRC used in the project. This RPC was developed under the Roads to Reuse Program that was managed in collaboration between Main Roads WA and the WA Department of Water and Environmental Regulation. However, this certification was only used for crushed recycled concrete. One of the objectives of this certification is to test the quality of PwRC in compliance with the state materials specifications. For other materials particularly recycled steel and slag the project required the supplier to provide eco-labelling certification that contained information that could be used in the organisation's assessments.

Furthermore, the use of PwRC was recognised through the ISC V.2 rating that was awarded to the project. In this case study, all four interviewees were aware of RPC and its functions, advantages, and disadvantages (Table 16). Overall, except for the designer, the representatives of the other three stakeholders indicated their support for using RPC in construction projects.

> 'It would give confidence for the supply of crushed recycled concrete to projects such as Tonkin Gap… [it helps] to make sure that the crushed recycled concrete meets our [organisation] requirements plus health and safety requirements' [C_3P_1].

> 'Long term is beneficial, and it does minimise a lot of the waste' [C_3P_2].

> 'We think that having stringent processes, testing and licencing to supply recycled products is required and necessary to maintain a high-quality standard which in turn builds confidence in the products which in turn will self-generate greater demand for the recycled products' [C_3P_4].

In terms of limitations, the respondents indicated some limitations that they were aware of. These included differences in the requirements of various RPCs (C_3P_1), adding cost to the project budget (C_3P_1- C_3P_2), low capabilities of current recyclers to be certified (C_3P_2), and unclear value and lower than expected certifiers' performance (C_3P_3).

> 'So, probably the limitation is, different certifiers will assess different aspects of the give that certification. So, you can't necessarily compare like-for-like between the two competing certifications.' [C_3P_1].

> 'In some instances, they're used as green branding which incurs higher costs' [C_3P_1].

> 'The demand for large infrastructure projects using this material is fairly low still, so there might be three or four main roads projects happening in Perth at the moment and there only be one or two using this material. If demand is low, there's only going to be two or three suppliers that go through getting certified to produce the material' [C_3P_2]

'In Western Australia, the two companies are urban resources and WA recycling, so they actually take waste material off projects and crush it and use it for this product. So, your main suppliers of limestone or crushed rock don't have that capability' [C_3P_2]

'I don't know what value they would actually add to be quite honest; I think from the materials I'm working with the specifications are pretty robust in what the supplier needs to meet. I think a certification body was to me just a blood-sucking organisation' [C_3P_3].

'In [a project] suddenly we had asbestos and we had an independent guy going out there, again, a very competent, probably a $200 an hour engineer going out there and scratching around in the stockpiles to see if he could find asbestos or taking samples and putting it under a microscope to find asbestos. He was classified as an independent certifier, and he did the best he could. Did he really add a lot of value? I am not sure.' [C_3P_3].

8.4.1.4. Case Study 4 – OneOneFive Hamilton Hill Project

In this case study, no RPC was used to verify the quality of PwRC. However, a series of tests was done to ensure the demolition waste is contaminant free, particularly the tests targeted the existence of asbestos in the waste resources. Furthermore, the application of PwRC in this project was recognised through the Urban Development Institute of Australia (UDIA) as part of the demonstration of sustainability in this project. Lastly, the project client had engaged the consultant engineers to ensure that the PwRC to be used in this project would have the quality required by relevant specifications.

As indicated in Table 16, only the designer of the project was aware of the existence of such a scheme. In response to the question about the application of RPC, both client and designer respondents indicated that they would support their utilisations in construction projects.

'I think that the move towards industry complying with standards around safety and quality control of product is actually really important. And it's something that we need to do, I think, to build confidence within the industry, certainly. So, you know, to actually remove that concern about risk about people buying a product that ends up being contaminated, or, you know, with asbestos, or whatever it might be. So, I think having that compliance processes are really important' [C_4P_3]

The design consultant of the project indicated that two limitations with some RPCs are the cost of the process and the level and approach of assessment.

> 'The trick is, how do you do it? So, it's robust and reliable, but it doesn't make the whole process unwieldy so becomes uneconomic. So, it's always, what's the right level of rigour? You want to minimise the risk, but you never eliminate risk, but you don't want to make it so expensive that no one's going to do it' [C_4P_3]

8.5. Advantages of Using RPC in the Building and Construction Sector

Those who supported the use of RPC indicated six benefits emerging from their application. These benefits, as illustrated in Figure 46, are strongly interconnected. For instance, by assuring safety and quality control, the purchasing organisations are able to reduce the risk of PwRC, which in turn generates demand for these materials, followed by minimising C&D waste disposal.

C_4P_3 made it clear that the RPC application is important for PwRC quality assurance and building confidence for purchasing PwRC. This participant noted that

> 'The move towards industry complying with standards around safety and quality control of the product is actually really important. it's something that we need to do, I think, to build confidence among end users within the industry, certainly. So, you know, to actually remove that concern about risk about people buying a product that ends up being contaminated with asbestos, or whatever it might be. So, I think having compliance processes is really important.'

The same line of reasoning was provided by the client [C_3P_1] and supplier [C_3P_4] of Case Study 3 who maintained that

> 'It would give confidence for the supply of crushed recycled concrete to projects such as Tonkin Gap… [it helps] to make sure that the crushed recycled concrete meets our [organisation] requirements plus health and safety requirements' [C_3P_1]; 'we think that having stringent processes,

testing and licencing to supply recycled products is required and necessary to maintain a high-quality standard which in turn builds confidence in the products which in turn will self-generate greater demand for the recycled products' [C_3P_4].

The participant representing the builder of this case study (C_3P_2) expressed the long-term benefits of PRC in the industry even more strongly:

'Long-term [it] is beneficial, and it does minimise a lot of the waste.'

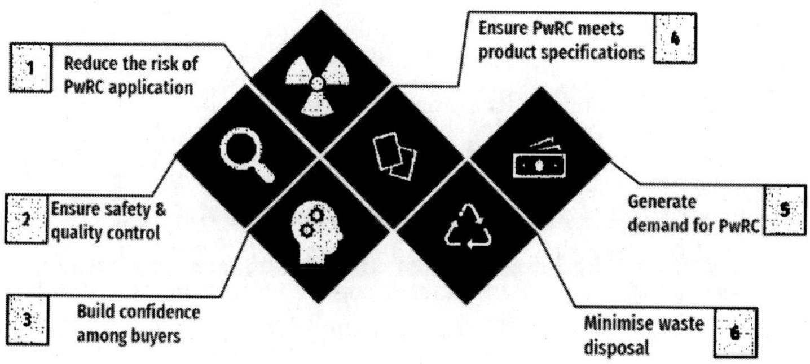

Figure 46. The main benefits of RPC application in the built environment sector.

8.5.1. Limitations of RPC in the building and construction sector

As indicated above, not all stakeholders expressed their support for using RPC in the sector (Figure 47). Even some of those who supported the RPC application or had limited to no prior knowledge of it referred to the limitations of relevant applications. As shown in Figure 47, stakeholders highlighted seven major limitations relating to various aspects of RPC applications. Three of these limitations were related to the schemes per se: the abundance of RPCs with different requirements *(Limit. #2)*, lack of transparency in their requirements *(Limit. #3)* and limited applicability for certain products *(Limit. #4)*.

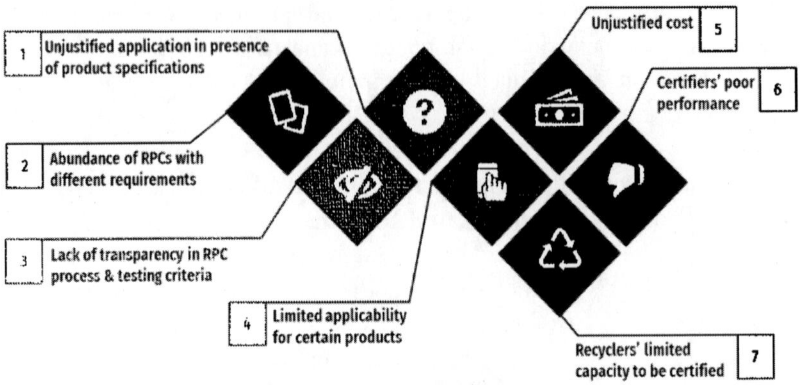

Figure 47. Major limitations of RPC application in the built environment sector

While supporting their application, C_1P_1 and C_3P_1 explicitly indicated that the abundance of RPCs with different requirements is a major barrier to the wide application of PRC. According to these participants, different requirements can be confusing and make it difficult to compare PwRC certified by different organisations. C_1P_1 stated that

> 'One of the challenges is that there are obviously so many different kinds of certifications.'

Although their organisation does not have a preference for a specific RPC, C_3P_1 highlighted that

> 'The limitation is, different certifiers will assess different aspects to give that certification. So, you can't necessarily compare like-for-like between the two competing certifications.'

Furthermore, the complex nature of some RPC programs and their untransparent process and testing criteria can be a barrier *(Limit. #3)*. C_1P_1 noted that

> 'Some of the certifications are a little bit opaque. So, you often find certification, you just get some sort of points or star rating system, but you don't really even know what that even means, you know, Level A or number three, four. It doesn't tell you anything, it just tells you that there's some sort of rating system.'

Another major barrier to RPC application was the certification cost being deemed unjustified *(Limit. #5)*. The builder's representative of Case Study 2, C_2P_2, noted that

> 'The problem would be if the product starts coming at a huge premium because of the certification process, well, that might be problematic.'

C_4P_3 maintained that

> 'The trick is, how do you do it? So, it's robust and reliable, but it doesn't make the whole process unwieldy so becomes uneconomic. So, it's always, what's the right level of rigour? You want to minimise the risk, but you never eliminate risk, but you don't want to make it so expensive that no one's going to do it.'

The low demand for PwRC making their application unreasonable was another reason. C_3P_2 stated that

> 'The demand for large infrastructure projects using this material is fairly low still, so there might be three or four MainRoads projects happening in Perth at the moment and there [might be] only be one or two using this material. If demand is low, there's only going to be two or three suppliers that go through getting certified to produce the material.'

The non-necessity of PRC in the presence of existing product specifications was another reason (Limit. #1). C_3P_2 mentioned that

> 'I don't know what value they would actually add to be quite honest; I think from the materials I'm working with; the specifications are pretty robust in what the supplier needs to meet. A certification body to me is just a blood-sucking organisation.' *C_2P_4* indicated that 'we have specifications for a reason. We need to maybe think about how we make those specifications more responsive, more agile would be far more useful than creating certification on top of certification to say that a product is potentially viable, that has nothing to do with the actual road building specifications.'

Lastly, C_1P_4, who represented the supplier that supplied recycled timber to Case Study 1, indicated that RPCs might be not suitable for some PwRC

(Limit. #4), such as timber, due to their physical nature. The participant added that

> 'It is a natural product, especially with timber there can be movement with timber, it's a live product, it does move so it moves to its surroundings and stuff like that.... in terms of certificates and certifications and stuff like that we could never warrant or give anything.'

8.5.2. Impact of Limitations on Realising RPC Benefits

This section explains how the seven limitations identified above can diminish the benefits of RPC that were reported by the research participants.

Limit. #1 - In the Australian built environment sector, when purchasing and applying PwRC, the common practice is to refer to state-developed product specifications [96]. Many infrastructure projects now procure PwRC that meet the requirements stipulated in these documents. However, as indicated in the case studies, compliance with product specifications may not be sufficient to prove PwRC quality, performance, environmental footprint and other criteria that would have been otherwise tested when RPC is applied (Benefit #2). As a result, it may be less likely that many end-users would be willing to purchase PwRC for their construction project if only product specifications are to be used (Benefit #5).

Limit. # 2 - The variation in the requirements embedded in different RPC programs can diminish the value of their application and be counterproductive in building confidence among the end-users of PwRC (Benefit #2). It is apparent that dealing with various requirements and testing criteria is a source of confusion for both producers and buyers and adds to the complexity inherited in the current PwRC supply chain ecosystem.

Limit. #3 - Lack of transparency in the RPC process may cause issues when the criteria of product specifications are to be met. This limitation is intertwined with a decreased trust in the organisation or individuals who are responsible for approving proposed PwRC. If the RPC process is unknown to these parties, the provided information on the demonstration of PwRC characteristics is unclear or there are uncertainties about whether or not the certified materials actually pass product specifications (Benefit #4), then they are less likely to approve RPC application in their projects.

Limit. #4 - At times, certifying certain PwRC may not be plausible due to various factors. In the short run, this means the market for RPC becomes limited to traditional PwRC for which there are established methods and procedures for testing and quality assurance. As a result, the application of unconventional non-certified PwRC is subject to various risks (Benefit #1), as there is no level of assurance of their safety and quality (Benefit #2) that can persuade some end-users to purchase them (Benefit #5).

Limit. #5 - As indicated previously, in some instances, RPC costs may seem unjustified (Limit. #5). This limitation can demotivate individuals operating in the construction materials supply chain to choose PwRC over readily available and sometimes cheaper brand-new materials. This pattern in the short-run results in less demand for purchasing PwRC in the sector (Benefit #5) which in turn triggers less waste recovery and more waste landfilling (Benefit #6).

Limit. #6 - Some PwRC certifiers' poor performance is directly translated to losing confidence in purchasing PwRC among buyers (Limit. #3). In particular, the lack of due diligence shown by certifiers during testing and control to ensure PwRC safety and quality (Benefit #2) will result in increased risk severity and the likelihood of the application of improperly certified PwRC in construction projects (Benefit #1).

Limit. #7 - The application of PwRC has yet to be fully embraced in the Australian built environment. Therefore, many recyclers are unable to be certified externally and/or produce PwRC as required by end-users or specified in some PRC programs. This means that they are at the potential risk of losing markets to brand-new materials that can better satisfy end-users' needs. In practice, this risk leads to lower demand for PwRC (Benefit #5) and will reduce waste recovery activities (Benefit #6).

8.6. Operation of RPC Issuing Organisation: A Case Study

8.6.1. Case Study Description

To analyse the operation of RPC organisations that issue RPC for PwRC, a case study approach was used. The case study selected for this analysis is an Australian-based organisation that has been in operation for the last 10 years. The organisation has no affiliation, financial interests or pecuniary involvement in the manufacturing sector. The GreenTag program offers end-product certifications that help material manufacturers demonstrate the fulfilment of requirements specified by different sustainability rating tools and product specifications. It currently provides three categories of certifications that deal with PwRC depending on the users' needs and requirements: LCARate, GreenRate, and McV Process (Figure 48).

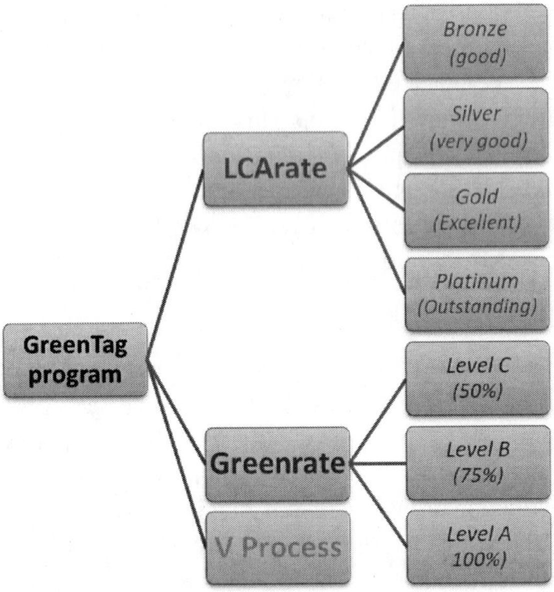

Figure 48. The three certification categories that are offered by the GreenTag program.

As mentioned previously, LCARate™ uses ISO-compliant lifecycle analysis and provides comparative LCA analysis of a product against a worst-case BaU product on a whole-of-life cycle basis. The findings of this LCA for this product determine its certification level: Bronze (good), Silver (very good), Gold (excellent) and Platinum (Outstanding). This assessment is completed through both remote and on-site audits.

The *GreenRate* certification covers green design, procurement, and facility management related products, and assists professionals working in these areas to achieve points under the GBCA's Green Star® rating tools and ISC's 'IS' rating tool credits [102].

With this certification, fitness for a purpose is assessed through the determination of the product's ability to meet the requirements of the product category specific national construction code (NCC) and GreenTag Standard requirements followed by receiving certification for that product. According to this assessment, the product can get three levels of certification: Level C (50%), Level B (75%) and Level A (100%). In Australia, New Zealand and

South Africa, this scheme is recognised for Green Star conformance assessment. It has been developed following the GBCA's Assessment Framework for Multi-Criteria Product Certification Schemes [103], 'Responsible Products Framework' [104] and is a GBCA-recognised product certification scheme [105].

McV process is the certification designed for the verification of manufacturers' claims for applying recycled content in their products. It uses the same principles as GreenRate but only focuses on PwRC. This option is specifically aligned with ISO 14021 self-declared environmental claims (Type II: environmental labelling) [106].

GreenTag also has a *Product Health Declaration (PHD)* certification scheme that relates peripherally to RPC in that it presents not only a *HealthRATE*™ 'In use healthiness' rating but a full transparency report on identifiable toxicological risks presents in products at an ingredient level. Given the unknown content of some recyclates, this can make such an assessment somewhat problematic.

8.6.2. Alignment with National ɔnd International Standards and Rating Tools

The GreenTag program overall is aligned and compliant with and/or certified under several national and international standards. Furthermore, certification under GreenTag can help any manufacturer demonstrate the sustainability criteria requirements specified by various sustainability rating tools such as Green Star and IS. As Global GreenTag is a trusted certifier, certifications issued by this organisation are also recognised by GBCA and ISC. Table 17 shows these standards, rating tools and their alignment with the GreenTag program.

Table 17. Relevant standards to which the GreenTag program conforms

Standard	Status/relevance to GreenTag Program	Context
ISO 9001 for Quality Management	Externally certified	Global
ISO 14024 for Type I (Third Party) Eco-labels	Externally verified & compliant	Global
ISO 14021 Environmental labels and declarations — Self-declared environmental claims (Type II environmental labelling)	Compliant	Global

Table 17. (Continued)

Standard	Status/relevance to GreenTag Program	Context
ISO 14025: Environmental labels and declarations — Type III environmental declarations	Externally verified & compliant	Global
ISO 17065 for Conformance Assessment Bodies	Externally verified & compliant	Global
ISO 14040 & ISO 14044 for LCA	Externally verified & compliant	Global
ISO 14067 for Greenhouse Gas calculation	Externally verified & compliant	Global
ISO 14025 for Environmental Product Declarations and	Externally verified & compliant	Global
ISO 21930 and EN 15804 for specific need EPDs	Externally verified & compliant	Global
Australian Competition and Consumer Commission (ACCC) Certification Trade Mark status	Certified by the legislator	Australia
US Trademarks & Patents Office Approved Trade Mark status.	Certified by the legislator	US
Sustainability rating tool	Organisation	Context
Green Star	GBCA, GBCSA, GBCNZ	Australia, Africa & New Zealand
IS	ISC	Australia & NZ
LEED	USGBC	Global
The WELL Building Standard™	International WELL Building Institute	Global

8.6.3. Findings from an Interview

One online interview was conducted to further understand the process of RPC for PwRC. The interviewee was the CEO of the organisation issuing RPC for C&D and other waste resources. The certifications provided through the GreenTag program are based on the data collected from remote and on-site audits, independent testing bodies, safety data sheets (SDS) and manufacturers' provided data and claims. The following sections show various characteristics of and issues around an Australian-based RPC.

8.6.3.1. Pricing Mechanism

The primary pricing mechanism in this organisation is a quote-by-quote basis. Despite having the price structure for any of the certification types, the cost of GreenTag certification is a function of various factors (Figure 49). These factors include the complexity of the assessment, the items to be assessed, the nature, mode (i.e., remote or on-site), the frequency of audits, and whether a third-party testing body is involved in the certification process.

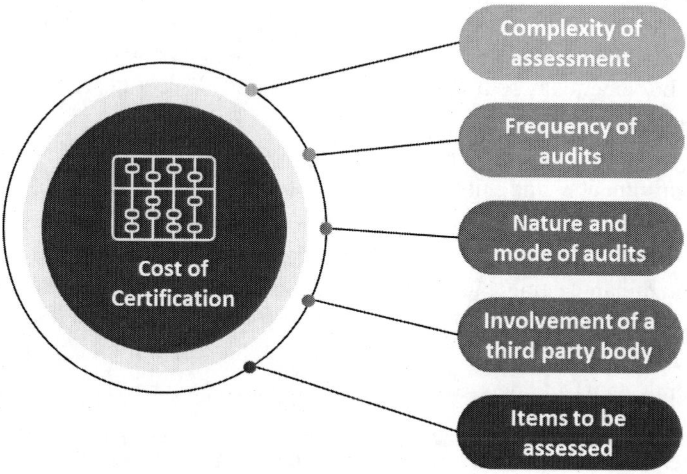

Figure 49. The factors that determine the cost of RPC.

According to the interviewees, among the factors described above, the inclusion of LCA has the most significant impact on the certification cost.

> 'Obviously, anything that involves detailed lifecycle analysis involves third-party contractors who do the LCA. So, any project that is wanting to do carbon mitigation to gain the carbon mitigation or the lifecycle impact mitigation credits in either of the rating tools will be asking for an EPD. Typically to date, Carbon data via EPD has been requested the big projects, but increasingly EPDs are becoming a default request on many projects.'

8.6.3.2. Process of Certification

While the certification is issued for the end product (Type I: ISO 14024 [107]), the manufacturing process is also assessed under the GreenTag program. Furthermore, all the GreenTag main certification options assess various aspects of the utilisation of recycled content in products that are to be certified. The items to be checked vary and can include durability, ethical supply chains, environmental performance, and chemical risks (health and ecotoxicity) such as chemical contamination.

'We don't yet do process certification. We do end-product certification. So, while we'll look at the process, and we'll make sure that the process is repeatable. So, we'll look at quality management processes and we'll look at consistency in the invoicing. We'll look at the quality of the inputs and the likelihood that the end product that we're certifying is likely to be the same today as it is next year, next month or next quarter.'

'We're also talking about ethical supply chains, we're talking about environmental and biodiversity outcomes, as well as chemical and potentially chemical contamination. If we were concerned that there were potential chemical contaminants, we'd be requiring separate chemical contaminant testing, but identify what chemicals were likely to be at risk and then have those raw materials tested for those risks. Often what we do is we also require the finished product to be tested for those same risk chemicals where identified.'

The pre-requisite assessment items are often developed according to the criteria stipulated in a sustainability rating tool, a product specification or an organisation procurement guidelines.

'That's the first thing we would do. We don't bother certifying something that isn't fit for purpose.'

The audits are done by the third party that is engaged by the manufacturers to do data collection as specified in the GreenTag program procedures. This third party otherwise known as an independent compliance assessment body should be ISO 17065 compliant or certified to do audits and collect the data as required. Depending on the materials' characteristics, a specialised testing organisation might be engaged.

'That will require the manufacturers to do that engagement, but we'll tell them what we need. If they haven't got it already, we would direct them to go and get the testing that we require… they have to be ISO 17025 certified. So, they have to be another independent testing organization that's independently certified and will trust that process… the testing body used would depend to a degree on what materials we were wanting to test, if we're testing plastics was different to testing soils or aggregates'

'If a manufacturer comes to us and says, we want to have you recognise the fact that this is suitable for use in Queensland, we would again say to them, well show us independent and third-party testing, that proves to us that your product meets the various state standards that you're wanting us to reflect in our certification'

The following figure shows some recent examples of PwRC that were certified by this organisation and used in the built environment sector (Figure 50).

Figure 50. Some examples of PwRC certified by the GreenTag program.

8.6.3.3. Major Stakeholders

The interviewee mentioned the main stakeholders who influence the utilisation of RPC as governments (at all levels), leading companies, clients and developers and the architect and design (A&D) community who are keen to use sustainability as their selling marketing point. Figure 51 summarises the main findings of the interview.

Figure 51. The summary of the main interview findings.

8.6.3.4. Drivers of RPC Uptake in the Australian Built Environment Sector

According to the interviewee, several factors will enhance the adoption of RPC in the Australian context. These include the increase of the awareness of circular economy, recent attempts to combat the impacts of climate change, improve biodiversity & nature positive outcomes and change in the Australian political landscape motivating the public agencies to enforce policies that require the use of PwRC. The latter has been reflected in the mandatory utilisation of PwRC in public projects in some Australian states.

> 'It's a growing necessity, particularly with the increasing awareness of circular economy and carbon and basically, its climate impact has had a really significant increase in the requests for EPDs in the last 12 months. I see that is growing again, now that there's been a change of heart politically and we don't see the systematic suppression of issues around climate change and the new focus on nature positive biodiversity outcomes at a governmental level, it will mean that all of the departments that really want to sort of do something and have the procurement policies already in place, will be able to enforce them'

In December 2020, the Department of Climate Change, Energy, the Environment and Water (DCEEW), released its first plan [108] to promote nature-positive concept and indicate the government's commitment to conserving the Australian nature. This plan guides Australian businesses to refine their activities in favour of nature.

8.6.3.5. Barriers to Uptaking RPC in the Australian Built Environment Sector

One of the issues that were raised by the interviewee about the adoption of RPC, was unfair criticism of these schemes that are made out of conflict with some people's commercial interests in the built environment sector. The second issue referenced by the interviewee was the fact that RPC is not among the top priorities at the moment.

> 'You've also got to be aware that there are segments of the industry that will say whatever they choose to say, to try and reduce people's reliance on certification and try to offset the effectiveness of certification. Because they don't want to have to go through the robust processes, that people who are certified go through, you know, these processes, they're not for the weak-hearted you know, there's a lot of work involved in getting these certifications, a lot of documentation, a lot of engagement.'

> 'It's about priorities because the cost of certification is a fraction of whatever their marketing budget might be in any single month, let alone a year. So, particularly for the bigger companies, it's not about cost, it's about priorities. If they can see that their certification becomes part of their marketing budget, then the cost of certification is, you know, single-digit percentages of most marketing budgets.'

8.6.3.6. Strategies to Improve the Adoption of RPC in the Built Environment Sector

Three strategies to increase RPC adoption were noted by the interviewee. These include the mandatory application, incorporating RPC cost in the marketing budget, change in policies in favour of the RPC application, make it a priority in rating tools. Particularly, the change in procurement policies of companies, governments and other agencies will develop a market for PwRC and RPC accordingly.

> 'As soon as they get a project, go to manufacturers and say, 'you're not going to get this job, unless we know what your certified recycled content is,' then they pay attention… pointing out to manufacturers that they're not going to get on the project, or they'll have a better chance of getting on the project if they're certified'

> 'So, particularly for the bigger companies, it's not about cost, it's about priorities. If they can see that their certification becomes part of their

marketing budget, then the cost of certification is likely, single-digit percentages of most marketing budgets'

'So, the first thing that needs to be engaged in the Government and corporate procurement policies that prioritise RPC that are already in place in the market need to be enforced, there's already you know, 60 plus billion dollars' worth of market spend out there looking for products that are certified'

'To have potentially to have the rating tools call it up as a renewed focus, it used to be a focus, it's not anymore as a specific icon.'

8.7. A Model to Determine RPC Application

8.7.1. Strategies to Address the Limitations of RPC Application

To optimise the use of RPC, the existing limitations need to be carefully addressed. Hence, this section focuses on the main practical strategies to serve this purpose. Overall, six strategies were identified that assist stakeholders in minimising the impact of limitations identified above. These strategies are described below:

Strategy #1 - Effective quality monitoring and certification of PwRC by suppliers are required to instil and maintain stakeholders' trust in the PwRC application. This must be accompanied by more vigorous government action in the form of legislation and standardisation [91]. Primarily, this strategy deals with three limitations *(i.e., Limit. #1, #2 and #3)*. The government regulations can drive RPC application in conjunction with current or future product specifications and set a regulatory assessment framework to proctor and standardise the operation of certifying organisations (Limit. #2 and #3). The latter can be assigned to specialised public agencies that can be established in light of supportive regulations.

Strategy #2 - Together with the industry associations, the government should also initiate promotional activities via the media to change the general public's attitudes and help them better understand the potential benefits and objectives of RPC. The end-users' improved understanding will result in the creation of demand for purchasing certified PwRC, reduce the cost of certification and thus address the Limit. #5. In previous research, this strategy was described as a facilitator of environmental behaviour motivating people to pay a premium for products labelled as green [25]. Furthermore, promotional activities can inform stakeholders of specific objectives of RPC that are different from product specifications *(Limit. #1)*.

Strategy #3 - The government, in collaboration with organisations running environmental sustainability rating systems, may also organise regular training opportunities to promote changes in industry stakeholders' attitudes and work practices [97]. This strategy can create opportunities for the waste recovery industry to overcome the challenges related to capacity building that is required to take advantage of RPC for higher PwRC sales (Limit. #7). Furthermore, education can result in the advancement of RPC programs to cover a wider range of PwRC *(Limit. #4)*.

Strategy #4 - Contractual obligation or other client requirements to use certified PwRC is another effective strategy for the broad application of RPC [90]. Currently, in the EU, it is illegal to use uncertified PwRC in construction projects [92], resulting in more frequent uptake of RPC in the sector across the European territory. According to a report by the WSP [90], the advantages of mandatory RPC include the eradication of rogue traders *(Limit. #6)*, greater confidence in the data provided for all parties *(Limit. #3)*, less room for manipulation and provision of the credibility required to make the RPC valuable [90].

Strategy #5 - refers to the availability of qualified material engineers at reasonable prices. According to anecdotal evidence, these professionals are scarce, highly in demand and expensive to employ. By educating more individuals to become qualified materials inspectors, not only will sustainable jobs be created, but audit costs will also be reduced. Therefore, this strategy can be considered to reduce the impact of *Limit. #5, 6 and 7*.

Strategy #6 - The final strategy addresses the simplicity of certification methods for PwRC. The waste recovery industry's interest in RPC will be piqued by certification processes that are simple yet effective and require a fair amount of administrative work [90]. This strategy specifically targets Limit. #3 by providing a better platform for improved end-user understanding of requirements, which organically leads to optimised uptake of RPC in the sector. Evidently, the classification of PwRC into easily comprehensible categories, together with correct certification, facilitates future client purchases [91], as they will be able to acquire a product suited to its intended use.

8.7.2. Interplay between Factors Influencing RPC Application

Figure 52 is a model that shows the interplay between the factors influencing RPC application in the building and construction sector. The information summarised in this model reveals how the factors analysed and proposed in this chapter impact each other. Furthermore, the model can serve different purposes (Figure 52). First, those in the public or private sector who are interested in establishing a new RPC program or refining their existing ones can use this information to develop a framework that guides their business models and practices.

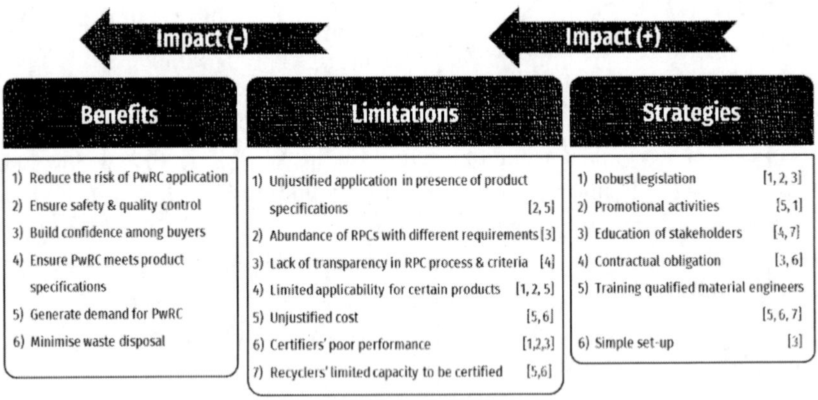

Figure 52. The interplay among the benefits, limitations and strategies for effective use of RPC.

Second, researchers and educators who focus on environmental sustainability in the built environment sector can use this information to further explore and communicate the role of RPC in the optimised uptake of PwRC in the sector. In Section 8.7, some directions for future research in this space are proposed.

Third, the policymakers who intend to embed RPC in procurement policies for purchasing PwRC can learn about the loopholes and strategies—and tackle them. In particular, it will assist them in developing informed policies that will reduce the risks of PwRC application and provide a level playing field for all stakeholders involved in the PwRC supply chain.

Fourth, as the application of RPC is aligned with the sustainability rating organisations' agenda, this information lends itself to reforming their sustainability assessment procedure that targets the utilisation of PwRC in construction projects.

Conclusion

In this chapter, the use of RPC was thoroughly investigated in three steps. First, the literature review was used to identify the main enablers and drivers of using RPC in construction projects. The advantages included (1) cost associated with their application (2) lack of cost justification (3) self-certification as an alternative, (4) lack of alignment between RPC and sustainability rating systems, (5) unreliable certifiers and (6) lack of collaboration between parties involved in certifying PwRC. The main enablers, according to the existing literature, are (1) robust legislation (2) promotional activities (3) education of stakeholders, (4) contractual obligation (5) training qualified material engineers, and (6) simple set up.

In the second step, the understanding of Australian stakeholders operating in the building and construction sector concerning RPC application was explored using a case study approach. To the best of the authors' knowledge, this study is the first of its kind in the Australian context reported in a scholarly attempt. This chapter offered useful information about the characteristics of RPC with a focus on the Australian context and generalisability to similar contextual conditions by analysing four case studies in which PwRC were utilised.

The key finding of this research is that, currently, the application of RPC is limited in the sector. Analysis of four case studies showed that only one out of four case projects applied this scheme. The chapter also listed the top seven limitations of RPC application, namely unjustified application in the presence of product specifications, the abundance of RPCs with different requirements, lack of transparency about their requirements, limited applicability for certain products, unjustified cost, certifiers' poor performance and, recyclers' limited capacity to be certified. To address these limitations, this chapter proposed several strategies aimed at various key stakeholders (government, end-users, industry associations, waste recyclers, certifiers, education providers and sustainability rating organisations) to make the RPC application business-as-usual, when proven useful.

Furthermore, the future research agenda should include an investigation into the effectiveness of the strategies provided in this research in reducing the identified limitations. Moreover, a cost-benefit analysis can be performed to estimate the certification cost under various scenarios. This will help stakeholders such as product suppliers and end-users obtain a deeper grasp of what they receive in exchange for the costs associated with PwRC certification. Such an attempt will eventually determine the usefulness of RPC application in the sector.

Chapter 9

Circular Economy Guidelines

9.1. Introduction

C&D waste generated during construction and infrastructure activities accounts for around one-third of global waste generation [109]. According to the latest national waste report (2022), Australia is no different as 29.0 million metric tonnes (Mt) of C&D waste was generated in 2020-2021, in Australia, which represents around 38% of Australian entire waste generation [9]. Around 6.4 Mt of C&D waste was disposed of in landfills. Through CE practices embedded with 'design for no waste' and 'design for reuse/recycling' principles, much of the C&D waste generation could have been avoided or at least recirculated within the construction and infrastructure supply chains. Since the China Waste Ban, major changes in the Australian regulatory policies have taken place (including Nation Waste Policy reform in 2018, setting up the action plan in 2019 and the 2025 waste recycling targets) in strengthening local waste infrastructure and improving overall waste management practices based on sustainability principles. The National Waste Policy Action Plan (1019) sets 7 targets, 17 strategies and 78 actions to implement the Australian Waste Policy and the following targets are highly relevant for C&D waste management:

- Reduce total waste generated in Australia by 10% per person.
- 80% average resource recovery rate from all waste streams.
- Significantly increase the use of recycled content by governments and industry by 2030.
- Make comprehensive, economy-wide and timely data publicly available to support better consumer, investment and policy decisions.

Like the Commonwealth Government, all states and territories in Australia have incorporated CE into their waste management strategies.

9.2. CE Policies in Different States and Territories in Australia

The CE strategies considered in the State and Territory Governments follow the core three principles (i) design out waste and pollution, (ii) keep products and materials in use, and (iii) regenerate natural systems. However, the application and practices of CE in the state and territory governments vary depending on their local context and priorities. For example, South Australia has a long history of driving sustainable reuse of waste through policy and legislation spanning more than 30 years (e.g., Container Deposit Scheme-CDS). Moreover, the SA Government has also initiatives to develop partnerships to implement the CE in regional areas [110]. NSW's current CE policies focus on seven key areas which emphasise the importance of 'push' and 'pull' initiatives to stimulate both the supply and demand sides of resources in the economy. The transition to a CE is enabled by data, investment, innovation, collaboration, skills and engagement [111].

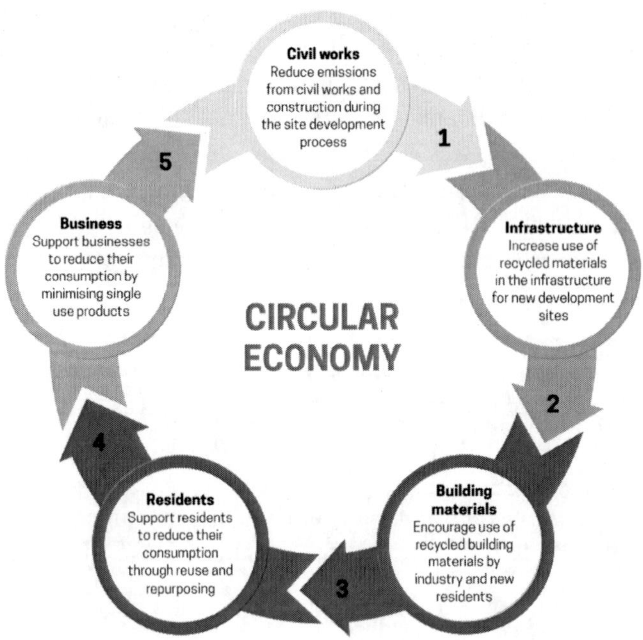

Source: ACT Government (2020) [113].

Figure 53. Five steps to achieving a CE.

Table 18. Summary of the States and Territory Governments' CE-related policies in Australia

States and Territories	Waste/CE strategy	Action plan/priority areas/Targets	Specific to C&D	C&D targets
Australian Capital Territory [116]	ACT Waste Management Strategy	Carbon neutral waste sector	No	No
New South Wales [111]	NSW Circular Economy Policy Statement- Too Good To Waste	Eight priority areas and three recycling targets	Yes	80% for construction and demolition waste
Northern Territory [117]	The Northern Territory CE Strategy 2022-2027	Three priority areas • Regulatory framework modernisation • Transiting to CE • Establish CE industry	No	No C&D related targets
Queensland [118]	Waste Management and Resource Recovery Strategy	Targets for 2050 • 25% reduction in household waste • 90% of waste is recovered and does not go to landfill • 75% recycling rates across all waste types	Yes	85% of C&D waste diversion to landfill by 2050
South Australia [119]	South Australia's Waste Strategy 2020-2025	By 2025: • 75% diversion of MSW • 90% diversion of C&I • 95% diversion of C&D	Yes	95% diversion of C&D by 2025
Tasmania [115]	Draft Waste Action Plan, 2019	• 10% waste reduction by 2030 • 80%resource recovery by 2030	Generic	No
Victoria [112]	Recycling Victoria A new economy	• Divert 80 per cent of waste from landfill by 2030 • Cut total waste generation by 15 per cent per capita by 2030	Generic	No
Western Australia [114]	Waste Avoidance and Resource Recovery Strategy 2030	• 30% reduction in C&D waste generation per capita by 2030 • Increase C&D material recovery to 80% • 85% waste diversion rate	Yes	80% material recovery target by 2030

The 'Recycling Victoria' is the Victorian Government's 10-year policy and action plan for waste and recycling with 26 action items to achieve 4 CE goals [112]:

- Goal 1 – Design to last, repair and recycle.
- Goal 2 – Use products to create more value.
- Goal 3 – Recycle more resources.
- Goal 4 – Reduce harm from waste and pollution.

The ACT Government aims to achieve CE by considering the following five steps, as shown in Figure 53. The ACT Government's CE strategies are highly relevant for C&D waste as they specifically consider emissions reduction, increased use of recycled contents in civil and infrastructure works and the use of recycled building materials.

The CE has been incorporated into WA's Waste Avoidance and Resource Recovery Strategy 2030 [114] and the Tasmanian Government's draft Waste Action Plan [115]. The NT Government has recently published their waste strategy, where the Government will work in partnership with local governments, industry, non-government organisations and communities towards the following three strategic priorities (Table 18).

- Priority 1: Modernise the regulatory framework to protect the environment and create the right regulatory settings for growing the CE industry
- Priority 2: Start transitioning the Territory to a CE.
- Priority 3: Establish the CE industry as a contributor to the Territory's $40 billion by 2030 vision.

Despite having a strong emphasis on CE principles in the national and state/territory governments' policies, clear guidelines for C&D waste management are yet to be developed for the building and construction sector in Australia. Therefore, this chapter reviews the issues related to "Design for No Waste" and "Design for Recycling" to help develop industry-specific CE guidelines. The review has a particular focus on the following points:

- The current CE practices in the construction and infrastructure projects in the global and Australian context;

- Key barriers and enablers of applying CE principles in the construction and infrastructure projects; and
- Industry-specific guidelines for implementing CE practices in the building and construction sector.

9.3. Trends of the CE Studies

This section of the chapter focuses on the trends of the CE studies in the building and construction context. As outlined in Table 19, Asian and European countries have the greatest number of academic publications on CE in this context.

Table 19. Geographical location of the selected articles

Region	Countries	No of Articles
Africa	Egypt and Nigeria	9
Americas	Brazil, Canada, and the USA	10
Asia	Indonesia, China, Hong Kong, India, Kazakhstan, Sri Lanka, Korea, Malaysia, Myanmar, Pakistan, Philippines, and South Korea.	35
Europe	Austria, Belgium, Finland, France, Italy, Germany, Poland, Luxembourg, Netherlands, UK, and Ukraine	46
Middle East	Iran, Oman, Saudi Arabia, UAE	3
Oceania	Australia	11
Global	Global	1

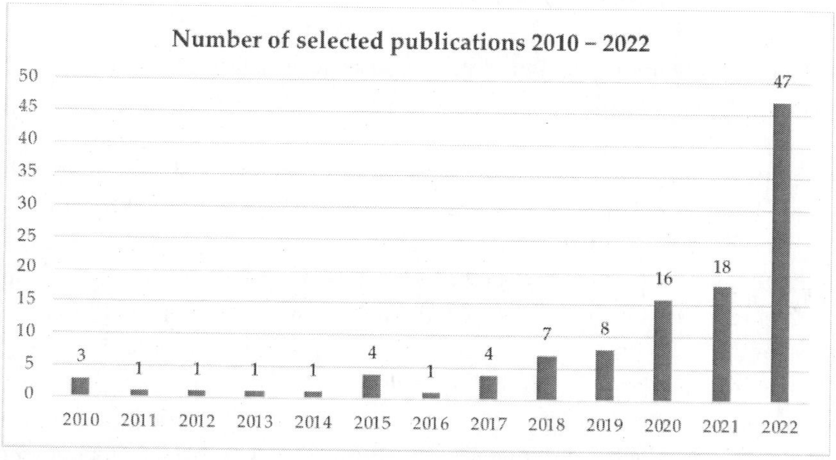

Figure 54. The number of articles published during 2020-2022.

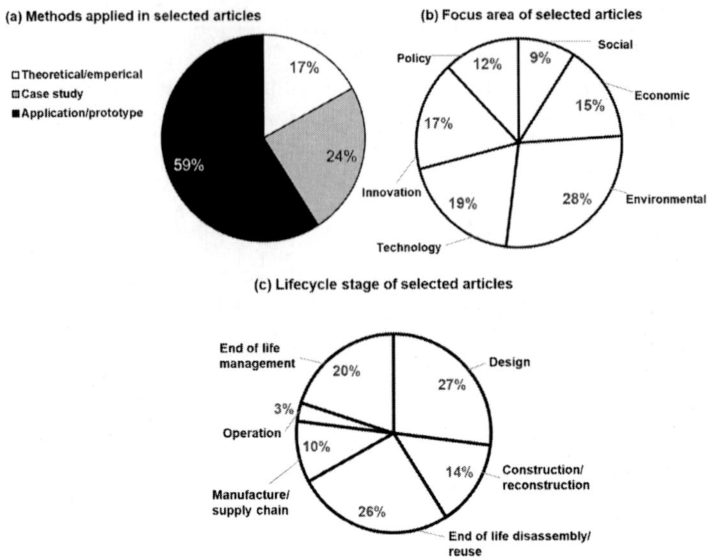

Figure 55. Statistics on the study's literature (a) research method, (b) focus area, and (c) life cycle stage.

The selected articles' data in the last 10 years shows that there has been an increase in the number of publications around the topic areas. Around 70% of the articles (81) were published in 2010 and 2022, showing an influx of recent research publications. In 2022 alone, the number of selected publications was about 2.5 higher than the year before (Figure 54).

Figure 55a shows that research on current CE practices in the building and construction sector has had a dominant theoretical focus rather than a practical one. Around two-thirds of the selected articles applied theoretical methodologies (literature reviews), while case studies and prototypes accounted for 22% and 13%, respectively. Figure 55b illustrates that technology, innovation, and environment are the most studied areas in the selected articles (between 22% and 23%). Less focus is given to economic (13%), policy (11%), and social (9%) aspects. Figure 55c gives an overview of the life cycle stage that the selected articles cover. The EoL stage makes up 42% of the research, with 22% addressing the EoL through processes such as disassembly and/or reuse, and 20% of the articles focus on the EoL from the waste management perspective. Nearly a quarter of the articles address the design stage, which is followed by both the construction and manufacture stages, at 18% and 15%, respectively. Interestingly, only 1% of the articles discuss the operation stage, showing a gap in research.

Table 20. The key areas and elements covered in the articles

Broader Scope	Specific Focus	Number of Articles	Key Elements Covered in the Articles
Circularity practices	CE/zero waste/no-waste	19	Zero waste and CE principles, understanding of key barriers and opportunities in the transition to a CE-built environment.
	Design for deconstruction/ disassembly (DfD)	12	The application of CE and zero waste design principles such as DfD or deconstruction. Design for reusability and adaptability and the opportunities for reusing the entire building by relocating and adaptive reuse.
	Reuse of building, elements, materials	11	The opportunities for reusing construction materials such as aggregates and timber, and reusing building elements/components such as steel frames and steel beams/columns.
Resource management	Recycling/Resource management	21	Waste resource management from the point of minimisation, recycling, and recovery of resources from waste, including post-disaster events and the relevant regulatory policies.
Innovation & optimisation	Digital technology	14	Digitisation and application of innovative technologies such as BIM, digital material bank, material passport, and cloud computing to enable data-driven decisions are the key focus of digital innovation.
	Construction innovation	13	The recent innovations in construction technologies such as off-site/modular construction, prefabrication, and green construction materials using alternative building materials sourced from recycled components seem to be the key areas covered by the articles.
	Project/supply chain optimisation	11	The project optimisation through innovative approaches such as the integration of aclient–designer interface and introducing an early contractor's involvement in the design phase to minimise waste, better manage materials, and optimise productivity.
Sustainability Assessment	Measurement	14	Measuring impacts of current practices and the sustainability benefits and performance in the context of sustainability priorities.

Based on the title, abstract, and keywords review of all 115 articles, the articles are grouped into 8 specific focus areas after completing multiple iterations of identification, selection, and regrouping of research focus and scope covered in the articles (Table 20).

9.4. Barriers and Enablers of CE Implementation in the Building and Construction Sector

9.4.1. Key Barriers

The application of the meta-analysis to the collected data served to identify the key barriers to implementing CE in the built environment sector in Australia (Figure 56). The results suggested four key clusters of barriers: cultural, regulatory, organisational, and knowledge and understanding. The literature examined in this section covers both Australian and other countries' sources.

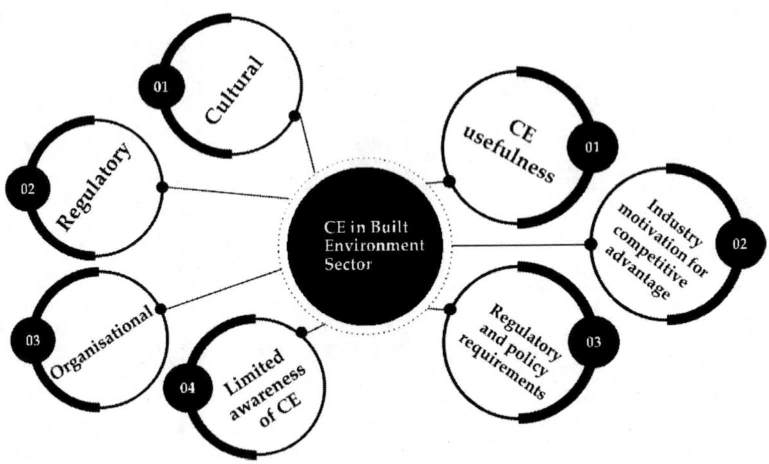

Figure 56. Categories of barriers and enablers of CE application in the built environment sector.

- *Cultural barriers* - There are some cultural aspects and social norms that heavily influence the building industry. First, the aesthetic preferences of designers, developers, and building occupants may be incongruent with some of the strategies to enable circular economies in the building sector [120]. Traditionally, embellishment of buildings shows wealth and/or status, particularly in residential settings. Secondly, another cultural barrier is the high-consumption lifestyles that have spread in many cities around the world. Lastly, professionals in the building sector tend to be resistant or slow to change and wary of incorporating new approaches [120].

- *Regulatory and policy barriers* - The legislation gives little attention to design phases which are essential to waste minimisation [121,122]. In Australia, there is a lack of supportive policies, regulations, and standards to foster circularity in the built environment [123]. Several issues impede the implementation of these essential regulatory measures. There are issues with the reclassification of waste as a resource, which hampers EoL treatment and material reuse. Additionally, there are inconsistencies in waste and materials management regulations across Australia, making it difficult to produce national frameworks [123].
- *Organisational barriers* - In Australia, some barriers at the organisation level inhibit the progress towards a more circular built environment, as was evidenced in a recent GBCA survey [123]. Demands such as emissions reduction or carbon footprint reductions take precedence. Likewise, the survey revealed that there is not enough stakeholder collaboration and a lack of agreed metrics and processing infrastructure.
- *Limited awareness and understanding of CE* - In Australia, there is little understanding of what a CE can represent to the building sector and the benefits it provides [122, 123]. In addition, there is a need to differentiate between a CE approach and a traditional sustainable model. The former is based on strategies that make business sense, while the latter has moved away from financial aspects. This can certainly discourage the uptake of a CE [124]. Overall, if there is no understanding of the value of a CE, there will not be a real drive to push the building industry into circularity [124]. A survey conducted on several industry stakeholders in Australia revealed that participants' awareness of the CE has a statistically significant impact on aspects including willingness to adopt a CE, CE adoptions, and benefits of the CE [122].

9.4.2. Key Enablers

From the selected data and meta-analysis, the following key enablers of the application of a CE in the built environment were identified. The three top enabling clusters are the aspiration for sustainability, competitive advantage/cost reduction, and regulatory requirements. The literature examined in this section covers both Australian and other countries' sources.

- *CE usefulness* - Research has shown that there are several driving factors related to sustainability that can enable a CE in the building sector. These include resource and energy conservation, enhancing indoor environmental quality, and waste minimisation [125]. Providing evidence of CE-added value has been identified as a major driver to implementing CE in the Australian built environment [122]. Industry motivation for competitive advantage/cost reduction-Investing in sustainable solutions may be seen as a cost premium in a project building, and there could be long payback periods before recovering the initial investment [120]. However, provided that 50%-60% of construction project costs account for materials, reducing waste is a cost-saving strategy [126]. This means that material waste heavily contributes to cost overrun. For instance, it has been shown that waste materials such as concrete, piles, and stone tablets increase project overrun by 7%, 13%, and 26%, respectively [126]. In short, not only do material cost savings improve financial performance, but they also can enable circularity in the built sector.
- *Regulatory and Policy Requirements* - The policy is essential to enable a CE in Australia's building sector. Stringent legislation and policy have been identified as key success factors in enabling DfD of buildings, which lends itself to designing out waste [121, 127]. Waste construction effectiveness is dependent on the extent to which it was designed out [121]. Governments must set building deconstruction targets and provide incentives to achieve them to reduce waste intensiveness. This is a proven way to enable DfD of buildings as construction works require planning acceptance and authorisation based on current regulatory frameworks [127]. Other policies that have been identified as (potential) waste minimisation enablers in the building and construction sector are [121]:
- *Economic carrots* - these could be in the form of tax reliefs or incentives. These measures are especially welcomed by construction

contractors and site workers, who are highly motivated by the economic benefits of waste minimisation [128]. The building and construction sector experts and workers would be incentivised (not coerced) to reduce waste. A study conducted in Australia showed that relevant stakeholders perceive the lack of incentives as an important obstacle to CE implementation in the building sector [122].

- *Sustainable construction appraisal systems*: most of these tools allocate points for material sorting, recycling, and/or reusing (e.g., US's LEEDS, UK's BREAAM). The same point allocation mechanism could be applied to waste management, which could be a key factor in designing out waste.

9.4.2.1. SWOT Analysis

The data collected from the selected literature were analysed to identify strengths, weaknesses, opportunities, and threats (SWOT). The SWOT analysis (Table 21) provides an initial understanding of all data collected.

Table 21. A SWOT matrix was developed to gain an initial overview of selected data

Strengths	Weaknesses
• Circularity in the built environment is gaining traction, as there are more documents studying this matter every year. • There is consistency among information from different sources; there are no contradictory recommendations and/or strategies as to how to achieve circularity in the built environment through design out waste (DoW).	• Theoretical methods are more popular than case studies or application approaches. • Many studies focused on the positive outcomes for the environment when applying circular methods; economic and social aspects are often disregarded.
Opportunities	**Threats**
• Few studies/documents mention the social contributions of promoting circular built environments. • Design and EoL life-cycle stages seem to have more circular strategies than other phases, namely, operation.	• There is hardly any information that supports or proves to the industry why applying circular practices is worthwhile, especially economic-wise. • Most cases of success have been trialled in other countries, with different policies and contexts to Australia.

Table 22. CE Guidelines for the building and construction sector and alignment with the EMF CE principles

Broader Scope	Strategy	KPI	Alignment with EMF CE Principles		
			DoW & Pollution	Keep Products & Materials in Use	Regenerate Natural Systems
Circularity practices	1. CE procurement	Both qualitative (yes/no) and quantitative (%) of the PwRC	**High**	**High**	Medium
	2. Design for Deconstruction (DfD)	Disassembly Potential Rating	**High**	Medium	Low
	3. Design for flexibility and adaptability	Adaptability and Flexigf	Medium	**High**	Low
	4. Design for long life	Both qualitative (yes/no) and quantitative whole life-cycle cost (WLCC) [$/m²/yr]	**High**	**High**	Low
	5. Eliminate building components	Material use intensity per functional unit (kg/unit/year)	**High**	Low	Low
	6. Reuse building/building elements	Reused floor area (% of total GFA) or building component reuse in (%)	**High**	**High**	Medium
	7. Restore and regenerate	Soil sealing factor and/or compensatory measures (rainwater management, green roofs)	Low	Low	**High**
	8. Design out hazardous/pollutant materials	Environmental Impact Cost [$/m²/year]	Medium	Low	**High**
	9. Climate resilient design	Embodied Carbon Intensity [$kgCO_2$ eq/m²/year]	Medium	Low	**High**
	10. Sharing economy/shared space	Provision of the shared economy (yes/no) or % of shared space	Medium	**High**	Low
Resource management	11. Waste prevention on the construction site	Diversion rate from landfill (%)	Medium	**High**	Low
	12. Material/component recycling	Proportion (mass/unit of reference) of secondary materials installed in the building. GFA could serve as the unit of reference	**High**	**High**	Low

Broader Scope	Strategy	KPI	Alignment with EMF CE Principles		
			DoW & Pollution	Keep Products & Materials in Use	Regenerate Natural Systems
Innovation and optimisation	13. Use of digital technology (e.g., material passports)	The proportion of building components or traceable materials in %	Medium	Medium	Low
	14. Construction Innovation (e.g., modular construction)	Proportion modular and/or off-site construction in %	Medium	Low	Medium
	15. Green supply chain (e.g., use of bio-based materials)	Local vs overseas/sustainably sourced (%)	**High**	Low	**High**
	16. Use and integration of sustainable technology	Energy demand from renewable sources (%), Energy storage capacity (kWh/time), microgrid options, etc.	**High**	Low	Medium

9.5. CE Guidelines for the Building and Construction Sector

Analysis of both scholarly and grey literature led to the development of guidelines for implementing CE practices in the building and construction sector. The guidelines were based on the broader scopes identified during the literature review. Table 22 shows an overview of the developed guidelines. Performance sustainability measurements in either qualitative or quantitative key performance indicators (KPIs) are provided for the strategies shown in Table 22. In addition, each strategy was evaluated using a heatmap according to the level of alignment with the Ellen McArthur Foundation (EMF) CE principles. The levels of alignment are high, medium, and low. A high alignment means that the strategy could contribute significantly to achieving the desired outcomes of each EMF CE principle, a medium alignment means a contribution to a certain extent, and a low level means a low contribution.

9.6. Circularity Practices, Resource Management and Innovation and Optimisation

The sections below describe the strategies in each of the three broader scopes identified in the literature review: circularity practices, resource management, and innovation and optimisation.

9.6.1. Circularity Practices

9.6.1.1. CE Procurement

A number of strategies have been developed to include sustainability aspects in the tender process. By integrating sustainability right from the tender phase, the decisions can take an integrated, holistic approach and no longer have to rely solely on economic factors [129]. Explicit tender requirements that target the impact on health and the environment would be integrated into the invitation to tender. Technical requirements, such as durability, ease of recycling, or ease of recovery, can also be included in this phase. Certification organisations such as the German Sustainable Building Council (DGNB) award tender requirements that explicitly request or recommend using secondary materials or reusing materials for mineral construction products. Toxic and hazardous materials must not be accepted during the procurement stage. Other targets identified may come as PwRC in the tender and/or construction stages.

Procurement can also be decisive in controlling waste by improving logistics. Waste-efficient procurement actions, including just-in-time delivery, improved supply chain collaboration, and reduced packaging [121]. These actions can avoid excessive packaging materials, inadequate materials storage, and double or poor handling, all of which are associated with construction waste [121]. Similarly, another successful action that can be taken during the procurement process is to include a contractual clause in which sub-contractors are held accountable for their waste disposal [121, 122]. A study found that in the Australian context, construction project managers consider incorporating waste management in the design phase more than effective on-site waste management [128].

9.6.1.2. Design for Deconstruction (DfD)

Deconstruction is the careful, piece-by-piece disassembly of buildings [127,130]. Deconstruction of a building is also known as selective demolition or disassembly [130]. The main goal is to maximise the potential reuse and recovery of a building's components and materials [127,130] and to prevent demolition at the EoL [127]. The benefits of conducting deconstruction processes outweigh the cost so long as the value of the building component is preserved when it reaches its end-of-life [127]. In fact, through DfD protocols, the value of a building can be retained as well as increased [131]. If buildings are designed for disassembly, a wide variety of end-of-life scenarios for the building's components and materials can be reached, such as reusing, recycling, relocation, and composting [132]. Table 23 shows actions that have been identified as key when conducting DfD.

9.6.1.3. Design for Flexibility and Adaptability

This design seeks to provide the building with multiple life cycles to make the most of resources and materials in terms of spatial and technical domains [134]. The former entails the ability to transform with low resource consumption (if any), and the latter involves structural and material alterations to reuse materials in the future [134]. Some specific actions include:

- Increase convertibility: Allow for changes in building use by designing the building envelope to allow for more than one use or to allow modifications in window size and spacing [134];

- Use standard, simple construction tools and technologies [130, 132, 135]
- Avoid bespoke/tailor-made solutions and complex building geometries [136].

Table 23. List of actions that support DfD

Actions	Key Reference/s
Allow parallel rather than sequential disassembly	[132]
Use lightweight materials to facilitate the easy handling of components	[127, 130, 132]
Size components to suit the proposed means of handling	[132]
Separate structure from cladding to allow changes to the building envelope	[132]
Provide access to all parts of the building that are to be disassembled	[132]
Arrange components in a hierarchy of access related to life spans	[131]
Use a modular system that is compatible with existing standards	[130, 132]
Reduce, simplify, and standardise connections.	[131, 132]
Provide a means of identification of components and assembly instructions	[131, 132]
Design using an open system that allows for structural alternatives	[127, 132]
Allow for disassembly at all scales, from materials to whole buildings	[132]
Logistics/manual of disassembly.	[132]
Avoid cast-in-place composite systems unless they are recyclable and reusable and do not cause negative environmental impacts	[127, 135, 133]
Avoid the use of joints and/or screws	[127, 130, 132]
Avoid the use of chemical connections (e.g., adhesives, coatings)	[127, 130, 131]
Avoid the use of hazardous materials and compounds	[127, 130, 131, 136]

9.6.1.4. Design for Longevity

A key objective is to keep the value of materials and resources as long as possible by intensifying their use [134]. The means to realise this is by extending the service life of buildings by (*i*) specifying durable components [120], (*ii*) avoiding the use of synthetic materials that do not allow refurbishment [120] , (*iii*) prioritising standardised, modular elements, (*iv*) maximising the durability of the building structure through careful selection, protection, and maintenance of components, (*v*) making use of Whole Life-Cycle Cost assessment (WLCC) as a design assessment tool, (*vi*) assembling in a systemic manner that is suitable for maintenance and allows for the possibility of replacements, and (*vii*) designing for longer service life [120].

9.6.1.5. Eliminate Building/Building Elements

This strategy is underpinned by dematerialisation principles, which seek to reduce materials or resource inputs in buildings while increasing performance and functionality [120]. These principles have many similarities with the concepts of eco-efficiency and material/resource efficiency that have been researched and applied in the construction sector. The concepts aim for economically competitive products that result from using materials and resource inputs efficiently. Dematerialisation considers all the environmental impacts (direct or indirect) of the life cycle of a product from the design stage. Several dematerialising strategies can be applied to a building project [120], as illustrated in Figure 57.

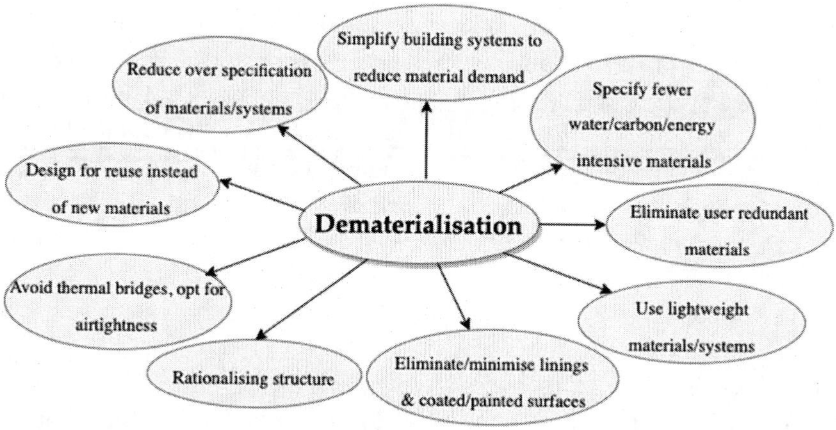

Source: Authors.

Figure 57. Summary of dematerialisation principles.

9.6.1.6. Reuse Building/Building Elements

According to the waste management hierarchy model [137], reusing waste resources is superior to recycling due to its environmental benefits, such as reduction in energy consumption and GHG emissions, the two major criticisms of using PwRC in the sector [22]. Furthermore, the technical advantage of reusing building elements is that they are reintroduced into the economy without requiring significant structural modifications or additional resources. Reuse can be performed either in the same or a new building location, and the function of the material or product may be different. Some examples of reusing building components include the reuse of greywater [138] and the adaptative reuse of existing structures [120]. Even though there are hurdles to overcome to reuse buildings or building components, which can be social, economic, material, or stakeholder related, the main obstacle is that they have not been designed for this purpose [132]. Therefore, design out waste (DoW) needs to be in alignment with sustainable manufacturing objectives.

9.6.1.7. Restore and Regenerate

This strategy aims to generate a positive impact on human and natural systems. Regenerative buildings repair, recreate and revitalise their sources of water, energy, air, waste, or any other matter [139]. Some of the actions include:

- Accommodating green spaces in and out of buildings to promote biodiversity and interaction between people and the surrounding nature. For example, by implementing green roofs, urban agriculture [134], trees, parks, gardens, green facades, or permaculture [139].
- Allowing high-quality indoor and outdoor environments to improve air quality. This could be achieved by implementing naturally lit/ventilated spaces or using vegetated walls or suspended gardens [139].
- Incorporating eco-friendly in-situ water treatment systems for both blackwater and greywater. The systems can be part of the greenery outlook and improve biodiversity [139].
- Soil sealing: Sparing use of land that minimises the impact on this land at a local level results in lower development, wastewater charges, and an improved microclimate [140].
- Implementing brownfield redevelopment processes when there is contaminated land to make it suitable for reuse. The existing land is significantly improved if the soil on the plot is properly disposed of. Soil can be deemed as contaminated not only when it is already classed as such but also if there is suspicion of contamination (polluting activities, spills, accidents) or if it contains munitions [141].

9.6.1.8. Design Out Hazardous/Pollutant Materials

Manufacturers must assess their materials according to the hazards of materials as well as the exposure routes during the use and EoL stages [139]. Materials that may be exposed to humans and nature should not contain harmful chemicals. If hazardous products are needed for any reason, they should be identified with a tag so they can be correctly handled at the end of their use [140,142].

9.6.1.9. Climate Resilient Design

The main goal of this strategy is to minimise the embodied carbon footprint of materials. The actions to realise this involve two main approaches:

- Using eco-friendly alternative materials: (a) bio-based products and finishing materials which can significantly reduce the embodied carbon of a building [36] (e.g., sustainably sourced timber by cement); (b) reused or recycled materials, preferably locally sourced (fewer transport emissions) (e.g., reclaimed bricks, locally recycled aggregates); (c) high-durability, low-maintenance materials (e.g., components with same lifespans as the building).
- Using fewer materials: (a) design optimisation leads to a reduction in quantities of materials used [29] (e.g., compact buildings that enable natural ventilation, lower wall/floor ratio); (b) design for deconstruction; (c) design for less on-site waste (e.g., off-site construction).

A good practice that can be implemented towards the achievement of this strategy is tracking and tracing the embodied carbon footprint of the building parts (e.g., structure, envelope, systems, fittings) and specifying a target below the recommended limits of the local area [33]. The use of digital technologies can facilitate this practice.

9.6.1.10. Sharing Economy/Shared Space

This strategy is concerned with intensifying the functional use of materials. This may involve reducing floor area, incorporating shared economy/functions, and designing for multi-functionality and adaptability [120, 134]. It must be stressed that none of these actions should hinder the building's capacity to meet basic user needs and/or operational performance [120].

9.6.2. Resource Management

9.6.2.1. Waste Prevention on the Construction Site

Innovative processes and concepts have paved the way for waste reduction at the construction site. Off-site construction considers the prefabrication of components, precasting of structural elements, and design for off-site construction [25]. By constructing off-site, waste-generating factors such as unexpected design changes, materials handling, and poor storage are completely prevented [25]. Indeed, ensuring a few design changes during the construction process has been identified as one of the top strategies for on-site waste minimisation. Provided that off-site construction helps reduce the number of design changes, this is an effective strategy to minimise on-site waste [121]. Other key strategies to combat waste in the construction site are:

- Following the project drawing/designs as closely as possible [142];
- Identifying potential problems to improve the quality of the final design [143]
- Assigning a dedicated place to collect and sort waste [142];
- Setting up waste bins in each building zone of the construction site [142];
- Optimising material waste segregation by providing skip bins for specific materials [142];
- Providing designers with technical information and capabilities of materials and equipment [143] ;
- Instigating synergistic interactions with other industries to recover the value of construction waste and by-products (e.g., Industrial Symbiosis where the waste of an industry is used as a resource in another industry) [144].

9.6.2.2. Material/Component Recycling

The use of secondary products may not necessarily contribute to ongoing project waste reduction, so contractors do not see it as an effective strategy to reduce construction waste. However, using PwRC is both essential to increasing waste diversion from landfills and decreasing waste intensiveness in the construction sector [142]. Using secondary materials also reduces the demand for raw materials extraction and their associated impacts [76]. Thus, PwRC should be used in construction projects. A self-declaration or manufacturer declaration specifying the material, component, or PwRC must be requested. Secondary raw materials contents can be either manufacturer-specific or sector-specific. Recycling approaches can also be considered when using aggregates in components (e.g., RCA) or when deciding on fixtures or fittings [120]. It is worth mentioning that recycling should only be practised if it saves energy and emissions compared to primary production and that it is technically feasible. Having said that, sometimes recycling leads to loss of quality, and sometimes there are no recycling routes for materials that retain their initial structure and quality (e.g., cement) [145].

9.6.3. Innovation and Optimisation

9.6.3.1. Use of Digital Technology

Although the built environment has not seen the same digital technology (DT) progress as other sectors, there have been important developments in the last decades [134]. These days, a variety of technologies can be used in different life cycle stages of a building. Some of the DTs include Artificial Intelligence (AI), Block Chain Technology (BCT), Building Information Modelling (BIM), Digital Twins (DTwins), and Material Passports and Databanks, among others. DTs can bring about several benefits to the building sector. Firstly, they foster supply chain collaboration among stakeholders as they can promote value networks, data-sharing, and knowledge. This is of the utmost importance as collaboration has been identified as a key factor in facilitating the CE in the building sector [134]. Secondly, they reduce the demand for raw materials by streamlining design processes, thus reducing waste generation. Thirdly, DTs help track and trace materials and resources throughout the entire lifespan of a building, which permits them to capture embodied value when they reach their end-of-life [134]. Overall, DTs help to virtually represent a built asset with all relevant building information, including material properties, maintenance schedules, (dis)assembly, and operation manuals. With DTs, buildings are seen as material banks, which creates new end-of-life possibilities for these materials rather than becoming waste.

9.6.3.2. Construction Innovation

Several methods that are different from traditional *in-situ* construction activities can be implemented to support a CE in the building sector. Non-conventional construction methods include standardised and modular designs and the repetition of similar designs in more than one component. Both methods rank as the most popular prefabrication strategies. Some of the benefits of prefabrication include higher quality of the final product, reduction of material demand and waste generation, and improved health and safety [146].

9.6.3.3. Green Supply Chain

In terms of bio-based materials, a decisive factor in selecting construction materials is quality and cost effectiveness. Organic-based or ecofriendly (sustainably sourced) materials are often overlooked as they are more labour-intensive and/or more expensive than traditional construction materials [147]. A study found that replacing traditional materials in large amounts with more eco-friendly ones can be considerably challenging, as the quality of the traditional materials is already well established, and replacement could be an issue factoring into sector and market trends [148]. Policies that incentivise the application of more sustainable materials are the key to changing the situation. More companies would be pushed to use greener construction materials [148]. Along with policies that drive the selection of sustainably sourced materials, there are some requirements that construction companies can demand from their materials suppliers:

- A raw materials list documenting origin, extraction, processing stages, and locations where the processing takes place.
- A corporate mission statement that supports the prevention of adverse environmental impacts from their raw material production process activities. Other social corporate statements, including the prevention of human rights abuses and corruption, should also be requested.
- Use and integration of sustainable technology

This strategy plans to capture economic value from sustainable and regenerative building technologies [134]. Specific actions include:

- Using clean/renewable sources of energy which should be locally generated [141];

- Applying Passive House or Minergie designs to ensure minimum energy consumption and maximum thermal comfort [139];
- Grid compatibility: Energy that is generated in the building (positive energy generation) or on its land from renewable energy sources is fed to the district/the area in the immediate vicinity [141]. Positive impact buildings have technologies that enable such transfer [134]

9.7. Supporting Framework Strategies

In addition to these strategies (Tables 24 and 25), our initial research findings in this area recommend the following to further improve the effectiveness of RPC in developing and stimulating of end-markets for recycled C&D waste products:

- A government agency should be established to control the quality and production of these materials
- Collaboration with sustainable building (construction) assessment schemes such as Green Star and Infrastructure Sustainability to acknowledge product certifications
- Stakeholder engagement through industry associations such as the Waste Management and Resource Recovery Association of Australia and the Australian Council of Recycling to enhance their capability to deal with RPC requirements in their policies and practices. Better stakeholder collaboration is key because the current fragmentation of the building and construction sector hinders the achievement of circularity in this sector [134].
- Enabling third-party organisations to develop PRC schemes that are agreeable to both government and industry clients
- Guidance on clear technical specifications or standards on the use of recycled aggregate for structural applications to enable recyclers to produce products complying with the industry standards
- Government support for research and development to further increase the quality of recycled materials and to raise public awareness of potential benefits

Table 24. CE cheat checklist for construction projects

Action	Implemented?
Tender requirements include/recommend the use of secondary materials or the reuse of materials	☐
Tender requirements do not accept toxic and hazardous materials	☐
Tender requirements request the use of recycled content (i.e. PwRC)	☐
There are waste-efficient procurement actions, such as just-in-time delivery and/or reduced packaging	☐
Tender requirements include a sub-contractual clause where sub-contractors are held accountable for the disposal of their waste	☐
Only cast-in-place systems that are recyclable and reusable are applied	☐
The use of joints and/or screws is avoided	☐
Chemical connections (e.g. adhesives, coatings) are avoided	☐
Hazardous materials and compounds are not used	☐
The design allows for any possible future changes (e.g. change in window size) – design envelope	☐
Use standard, simple construction tools and technologies	☐
Bespoke/tailor-made solutions and complex building geometries are avoided	☐
Durable components have been specified	☐
Standardised, modular elements have been prioritised	☐
Parallel disassembly design	☐
Use lightweight materials that facilitate easy handling of components	☐
Components are sized to suit the proposed means of handling	☐
The structure is separated from cladding to allow changes to the building envelope	☐
There is access to all parts of the building that are to be disassembled	☐
Components have been arranged in a hierarch of access related to life spans	☐
Modular system that is compatible with existing standards	☐
Connections have been reduced, simplified, and/or standardised	☐
Provide a means of identification of components and assembly instructions	☐
Design using an open system that allows for structural alternatives.	☐
The design allows for disassembly at all scales from materials to whole buildings	☐
There is a logistics/manual of disassembly available	☐

Action	Implemented?
There are multi-use actions that increase building utilisation (e.g. movable internal walls)	☐
Technologies and tools are standard and simple, not complex	☐
Adaptability manual available	☐
Use of WLCC as design assessment tool	☐
Assembly allows for maintenance and for the possibility of replacements	
Reduce overspecification of materials/systems or simplify building systems to reduce material demand	☐
Eliminate materials that may be redundant and may not capture user needs	☐
Building systems are simplified to reduce material demand	
Specify fewer water/carbon/energy intensive materials	☐
Design for reuse instead of new materials	☐
Thermal bridges are avoided, airtightness is supported	☐
Structured is rationalised	☐
Materials and/or systems are not over specified	☐
Elimination or minimisation of linings and coated/painted surfaces	☐
Use lightweight materials/systems	☐
Reuse waste sources (e.g. reuse of grey waters, adaptative reuse of existing structures)	☐
Design by considering reduction in floor area	☐
Design for multi-functional spaces	☐
Green spaces are accommodated (e.g. trees, parks, gardens)	☐
Environments with high quality of air are promoted (e.g. implement naturally lit/ventilated spaces)	☐
Eco-friendly in-situ water treatment systems for both blackwater and greywater are incorporated	☐
Spare use of land	☐
Brownfield redevelopment processes are in place when there is contaminated land	☐
All materials, including hazardous (if any) are clearly identified	☐
Use of eco-friendly alternative materials	☐
Use of fewer materials	☐
Incorporate shared economy/functions	☐

Table 24. (Continued)

Action	Implemented?
Off-site construction whenever possible	☐
Following the project drawing/designs as closely as possible	☐
Assigning a dedicated place to collect and sort waste	☐
Setting up waste bins in each building zone of the construction site	☐
Designers are provided with technical information and capabilities of materials and equipment	☐
Instigating synergistic interactions with other industries to recover the value of construction waste and by-products	☐
Digital technologies are used	☐
Prefabrication strategies are used	☐
Materials suppliers have provided evidence of materials being sustainably sourced (e.g., origin, extraction)	☐
The project uses clean/renewable sources of energy	☐
The project uses designs that ensue minimun energy consumption (e.g., Passive House or Minergie)	☐
The project offers grid compatibility	☐

Table 25. Checklist when choosing materials that support a CE in construction projects

Question	Y/N/Value
Is it a fit-for-purpose material?	☐
Is the material easily available?	☐
Can the material be extracted locally?	☐
How much energy is required during the material manufacturing	☐
How much water is required during the material manufacturing	☐
How much waste is generated from producing the material	☐
Is there any negative impact during the production process (e.g., pollution)	☐
Can the material be recycled with local infrastructure?	☐

Appendix: Meta-analysis of the CE Practices in the Selected Articles According to Their Key Focus Areas

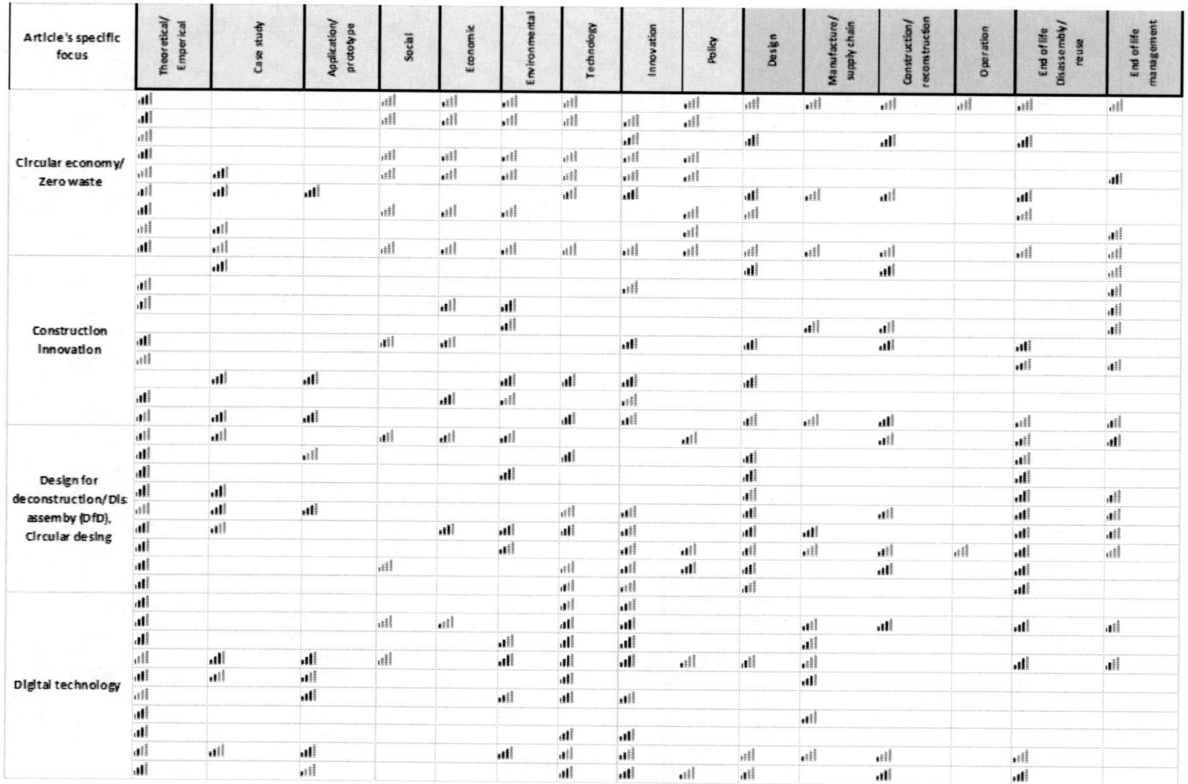

Appendix. (Continued)

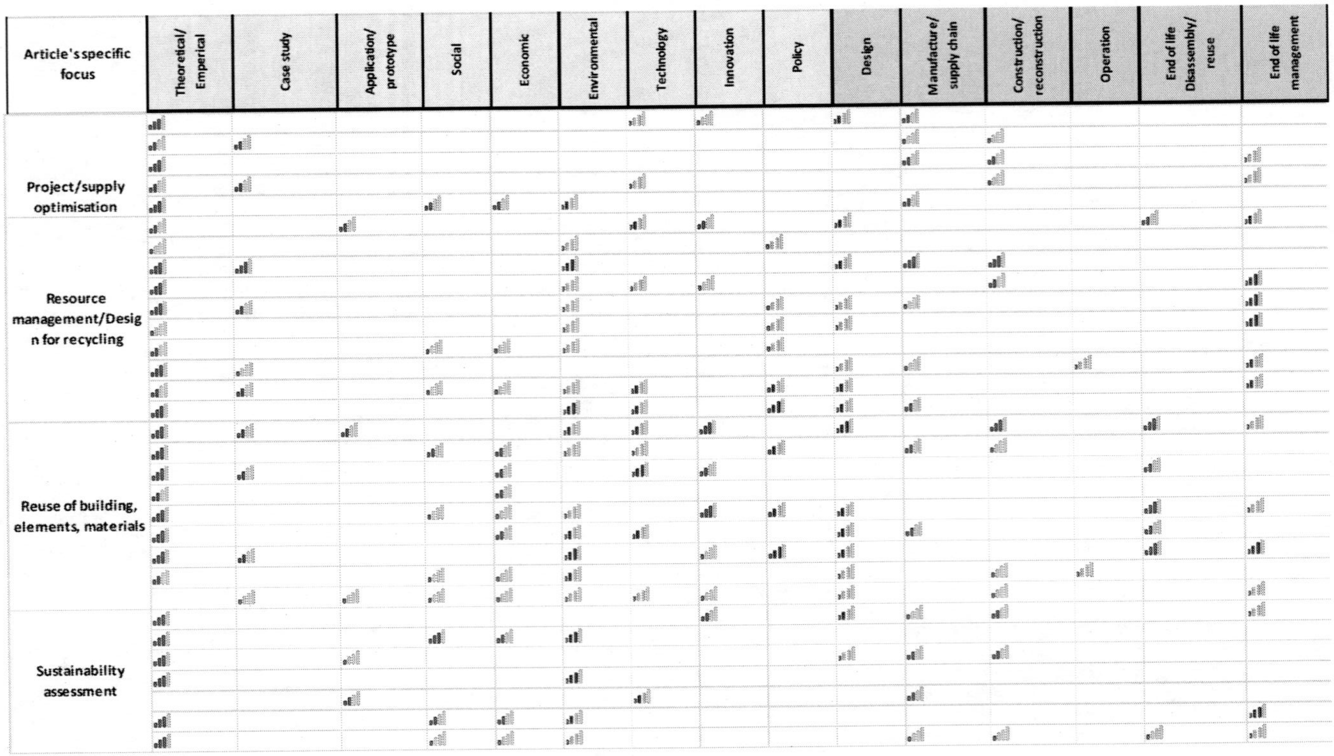

Conclusion

This chapter set out to develop CE guidelines for the C&D waste management of the Australian built environment sector. Although various design methods exist to minimise C&D waste, DoW guidelines have yet to be clearly outlined. This chapter sets a framework for the successful implementation of these guidelines. In doing so, a descriptive literature review was conducted to understand the current global and Australian CE practices in the sector.

This chapter also identified the main clusters of barriers and enablers of CE implementation in the built environment sector. Analysing these factors will help decision-makers make informed decisions about the extent to which CE principles can be applied in the sector. Finally, this chapter outlined industry-specific guidelines for implementing CE practices. Below are the main takeaways from these guidelines that will have sustainability management implications in the built environment sector:

- The guidelines present 16 strategies with detailed actions that apply to all stages in the value chain, from design, manufacture, and operation, to EoL. The project managers in the sector can adapt and apply these strategies where possible.
- The strategies encourage the sector to move away from the traditional, linear way of managing the built environment by keeping construction materials in the loop and improving waste management performance and using the built environment to regenerate nature (e.g., climate-resilient design). The latter is particularly important in delivering future projects where the demonstration of achieving (carbon) emission reductions through climate-resilient design is necessary.
- The seven categories identified as having a literature-specific focus (CE/Zero waste, Construction Innovation, DfD/Circular design, Digital Technology, Project/Supply Optimisation, Resource Management, Reuse, and Sustainability Assessment) served as a foundation to develop the three main scope areas of the proposed guidelines (circularity practices, resource management, and innovation and optimisation). Paying attention to these scope areas is necessary from the project management perspective to deal with barriers they face throughout the process of transforming to circularity.

- The circularity practices have a strong focus on various design concepts, namely, DfD, design for flexibility and adaptability, design for long life, design out hazardous/pollutant materials, and climate-resilient design. Resource management strategies are relevant when the material reaches its EoL. Innovation and optimisation strategies look to make the most of technological breakthroughs to realise circular built environments. Hence, designers and project managers in the sector need continuous education and training on the latest technological advancements to ensure that circularity will be business as usual.
- The literature review uncovered targeted strategies and actions that contribute to the DoW and pollution and keeping resources in use (two of the EMF CE principles)

Most notably, the CE guidelines for the building and construction sector suggested in this chapter should be seen as a dynamic framework that should be updated and revised with new research, good practices, technologies, and continuous learning.

Chapter 10

The Legacy Living Lab (Case Study 5)

10.1. Introduction

The importance of fostering decarbonisation and achieving global net-zero emissions is underscored by the escalating consequences of climate change, which pose both ecological and socio-economic risks [148]. Achieving the target of limiting global warming to 1.5°C above pre-industrial levels necessitates a coordinated effort at an unprecedented scale, involving governments, industries, and individuals [149]. Amongst various industries, the building and construction sector plays a significant role in global greenhouse gas emissions, making adopting low-to-zero emission building products crucial for sustainable development. Figure 58 shows that, in 2021, the buildings and construction sector accounted for around 37% of energy- and process-related CO_2 emissions and over 34% of energy demand globally [150].

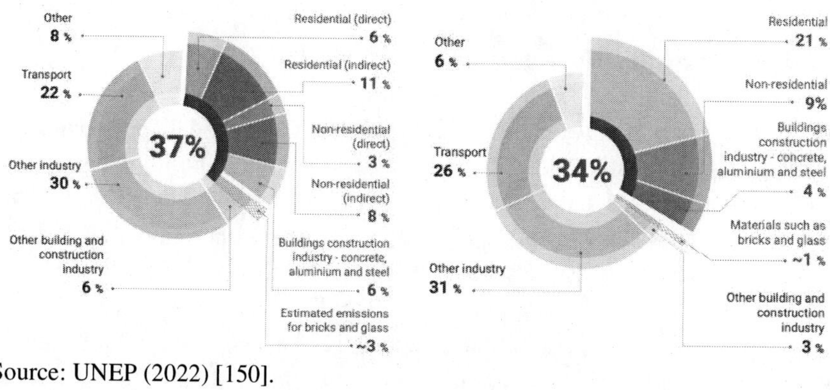

Source: UNEP (2022) [150].

Figure 58. Global share of CO_2 emissions (left) and final energy demand, 2021 by the building and construction sector.

Multiple factors contribute to elevated levels of carbon emissions in industrial activities. These include the types and quantities of construction materials used, construction methods employed, and significant waste generated during both construction and demolition activities. These elements also extend to the operational phase of buildings and infrastructure [151, 152].

The concept of a CE and Design for Deconstruction (DfD), based on modular construction, is emerging as a progressive alternative for mitigating climate change [142]. DfD encourages a paradigm shift in design, aiming for structures that can be efficiently disassembled at the end of their life cycles. This approach facilitates the reuse, refurbishment, or recycling of components [153, 154]. Consequently, DfD aligns closely with several pillars of the circular economy, such as waste reduction, resource efficiency, and overall sustainability, challenging the conventional "take, make, and dispose" model [155, 156, 157].

Applying the principles of a circular economy and DfD presents opportunities and challenges, as highlighted in various studies [158, 159, 160, 161]. For instance, Akhimien et al. (2021) [158] identify key obstacles, such as the absence of stringent legislation, inadequate information during the design phase, and a limited market for recovered components.

This chapter seeks to examine the Legacy Living Lab (L3) project by Curtin University in Perth, Australia—a pioneering example of DfD and the circular economy in action. The project aims for a versatile, adaptable space that can be efficiently deconstructed and reassembled [51].

10.2. Case Study Background

The L3 serves both as a visualisation centre and as a model for housing that displays unconventional construction materials, modular building techniques, and the contributions to the circular economy from regions such as White Gum Valley and the RENeW Nexus (Figure 59). The first operational cycle of L3 is slated to kick off in 2019 at the East Village location in Knutsford [162].

The L3 is a circular economy structure intentionally constructed for disassembly, deconstruction, and potential relocation. Additionally, the L3 serves as an innovation hub where the building and construction sector collaborates with research and the community to experiment, trial, and jointly create new building products and solutions. L3 offers a viable blueprint for how DfD can become standard practice in construction, serving as a nexus between academia, governance, and industry to push the circular economy agenda forward [162].

Figure 59. The Legacy Living Lab (L3) after construction.

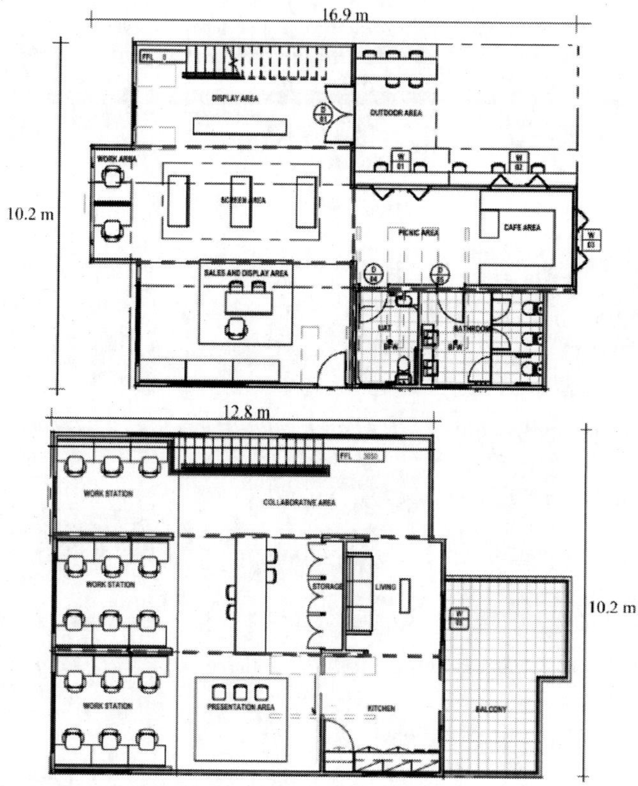

Source: Minunno et al. (2020) [161].

Figure 60. Detachable to disassembly component of the L3 as shown in the layout plan.

10.3. Design, Materials, and Construction Methods

L3 features a modular design, where each component is easily replaceable or reusable. Materials such as timber, steel, and bio-based materials were carefully selected based on their ability to be recycled and cause minimal environmental impact. The L3 project aims to showcase construction innovation through demonstration (Figure 60).

It is an excellent practical example for students to learn from. Moreover, it offers various industry partners to trial their materials and technology in a real-world setting, including dynamic blinds, smart appliances, smart indoor environmental monitoring systems, data visualisation and data centres etc. [162]. Fleetwood was the construction partner of L3. Figure 61 shows the key features and assembly of the L3 at the site.

(a) Modular construction — *(b) Joist joint - can be disassembled* — *(c) Screw joint for disassembly*

(d) Megantic ceiling and recycled carpet — *(e) Paver made from recycled plastics* — *(f) Indoor air quality mornitoring*

(g) Solar battery storage — *(h) Car charging station* — *(i) Rainwater harvesting*

Source: (https://l3.curtin.edu.au/gallery/).

Figure 61. Selected features of the Legacy Living Lab.

L3 applies circular building design, which entails adaptability, flexibility, and deconstruction at its core [163]. Instead of the traditional onsite construction approach, L3 applied offsite modular construction where the entire building components were made in the Fleetwood facility within the controlled and safe working environment [164].

Although a 1,200 mm framework of cladding materials was used, they were segmented into three pieces to reduce the managed weight from 60.6 kg to 20.2 kg to ensure safe handling for an individual and easy disassembly at the end-of-life phase [164]. According to O'Grady et al., (2021) [164], L3 is constructed primarily from steel, which comprises 61% of the structure's mass. Timber follows at 27%, with insulation materials making up 7%, and miscellaneous items like plasterboard and windows accounting for the remaining 5%. The total weight was 35.6 tonnes.

Although the priority was given to using PwRC and reuse once disassembly; however, new materials were also used, and some components needed to be recycled. For example, the external steel facade and vinyl flooring, cannot be reused.

The building components were transported and installed on-site with a joist joint instead of a welded joint (shown in Figures 60a and 60b). To ensure disassembly at the end of its life cycle, L3 features include visible and easy-to-access joints, limited use of welded connections, flexible design and disassembly options as shown in Figures 60c, 60d and 60e.

L3 also features smart technologies and monitoring systems, including the indoor air quality monitoring system (as shown in 3f), which has been proven to be very important during COVID-19. In addition, the building considers sustainable and renewable resource use during its operation, including solar power battery storage (as shown in 3g), electric car charging stations (as shown in 3h) and rainwater storage systems (as shown in 3i) to reduce its ecological footprint as low as possible.

10.4. Application of CE Principles in L3

The focus of this section is to analyse the L3 under the lenses of the 16 strategies identified in the guidelines for the Development of the circular economy guidelines for "Design for No Waste" and "Design for Recycling" (Chapter 8). Each strategy applied in the design, construction and current use of L3 was evaluated based on their level of consideration such as no consideration (NC), partial consideration (PC) and significant consideration

(SC) and how they were applied in L3. Table 26 shows the identified key CE principles and their application in L3. Table 26. The qualitative analysis of the identified circular economy principles being applied in L3.

10.5. Benefits and Outcomes

One of the key benefits of L3 is it was designed for deconstruction, meaning, the building components can be disassembled and refurbished easily for any adaptive purpose. O'Grady et al., (2021) [164] claimed that around 83% of walls and ceilings can be disassembled during the operation phase for any refurbishment purpose and the prototype building can be relocated, which in the process produces less than 1% of the buildings mass in waste. Apart from the disassembly and adaptive reuse of the space, there are various environmental benefits of the innovative approach. Minunno et al., (2020) [161] has conducted the life cycle assessment of L3 to examine the environmental impact from the construction of the facility. The study showed that compared to recycling, designing and building for reuse components offsets greenhouse gas emissions by 88%.

Conclusion

In line with previous research [156, 159], the L3 encountered various challenges, including material sourcing, joint fixing, and the absence of design standards tailored to Australian contexts [160, 164]. Despite prioritizing CE principles, L3 did not achieve full circularity.

A significant takeaway is the imperative for cross-sectoral collaboration. The project highlighted that effective DfD implementation is a collaborative venture involving design architects, developers, contractors, and policymakers. The modular approach has wide-ranging applicability, especially in temporary housing and disaster relief, underscoring the flexibility inherent in DfD principles.

The challenges predominantly stemmed from regulatory barriers, including intricate building codes. Conversely, L3 opened avenues for academic inquiry, community engagement, and commercial alliances.

Table 26. The qualitative analysis of the identified circular economy principles being applied in L3.

No	Circular economy principles	Application of the principles in L3 (No, Partial, Significant Consideration)	Application method
1	CE procurement	Yes- partial consideration	Prioritise PwRC being used
2	Design for deconstruction (DfD)	Yes- significant consideration	L3 was designed for demonstration in principle
3	Design for flexibility and adaptability	Yes- partial consideration	As modular construction, L3 has the flexibility for future adaptability
4	Design for long life	No	As a pilot living lab for research and development, L3 has been planned for short duration; however, the building materials are planned to use 2nd life application
5	Eliminate building components	Yes- partial consideration	Traditional concrete footing and foundation have been eliminated
6	Reuse building/building elements	Yes- partial consideration	The building elements had the possible application for reuse after disassembly
7	Restore and regenerate	Yes- partial consideration	Micro-footing and steel frame foundation help keep the soil in its original form and the land can be used for another purpose after disassembly
8	Design out hazardous/ pollutant materials	Yes- partial consideration	L3 use environmentally friendly materials as much as practically possible for the local context
9	Climate resilient design	Yes- significantly	The key focus was given on the climate resilient design by integrating water, energy and waste nexus during the design and operation of the building
10	Sharing economy/shared space	Yes- partial consideration	L3 is used as a collaborative space for startup companies and industry partners
11	Waste prevention on the construction site	Yes- significantly	Modular construction significantly reduced the waste creation during the manufacturing and assembly of L3 on-site
12	Material/component recycling	Yes- significantly	Recycled materials are used in L3 wherever practically possible

Table 26. (Continued)

No	Circular economy principles	Application of the principles in L3 (No, Partial, Significant Consideration)	Application method
13	Use of digital technology (e.g., material passports)	Yes- partial consideration	As a living lab a detailed material record was used to compare the assembly and disassembly at the end of the project life ((3-5 years)
14	Construction Innovation (e.g., modular construction)	Yes- significantly	As one of the innovative living lab, L3 applied innovative construction approaches as well as innovative technologies including smart sensors, air monitoring system
15	Green supply chain (e.g., use of bio-based materials)	Yes- partial consideration	As a multi-stakeholder' research project, the green supply chain couldn't be ensured in its entirety. However, the focus was given to environmentally friendly materials such as timber, recycled products and so on
16	Use and integration of sustainable technology	Yes- significantly	Various smart technologies were used and tested in L3, including smart blinds, smart air quality monitoring system, smart-data visualisation etc.

L3 serves as a pioneering example of construction aligned with CE principles. The substantial environmental benefits present compelling reasons for adopting DfD in future endeavours. As a living lab, L3's economic feasibility was not assessed, constituting a major limitation. Nevertheless, the project illuminated critical technical and constructability aspects of DfD within a local setting. Given L3's positive impact and insights gained, there exists a compelling case for the wider integration of DfD in multiple sectors. Future research should aim to alleviate regulatory barriers and promote industry-wide standards.

Chapter 11

Educational Programs for C&D Waste Management

11.1. Introduction

In response to C&D waste issues, three broad clusters of mitigation strategies, including enforcement, education, and encouragement, are offered. Education may play a crucial part in the 3R approach to waste management, which includes waste reduction, reuse, and recycling. Education has a significant impact on four primary domains in favour of waste management: increasing the awareness of industry stakeholders, improving organisational culture and changing their attitudes and behaviours. Furthermore, recent efforts to implement the CE concept in the building and construction sector place stress on education.

In Australia, the National Waste Policy [165] Strategy 3 (knowledge sharing, education and behaviour change), it is advised that coordinated knowledge-sharing and education initiatives should be implemented to address the needs of governments, businesses and individuals, and to encourage the redesign, reuse, repair, resource recovery, recycling and reprocessing of products. Most Australian jurisdictions have formed teams dedicated to educating industry people and the wider community. Research and education are critical factors highlighted in all Australian waste strategies and guideline documents. These documents have introduced initiatives to fund education programs to reduce waste disposal [78].

A report by the Senate Environment and Communications References Committee (2018) [166] recommended that the Australian government should support state and local governments in providing effective recycling education initiatives. There is an ongoing debate about the individual role of education in promoting sustainable behaviour and attitudes in this context. For instance, Crocker and Lehmann (2013) [167] highlight, that education alone has little or no effect when it comes to sustainable behaviour, concluding that improving knowledge and changing attitudes may not impact behaviour. On the other hand, Chapman et al. (2013) [168] observed that experiential learning results in new habits, and actively experiencing things is the most effective way to learn about environmental initiatives.

Several studies have highlighted the critical role of education in the C&D waste context. Results of a survey showed that 'culture, poor education (attitude and behaviour) and acceptance' is among the top four factors hindering a successful C&D waste management system [74]. In another research focusing on stakeholders' knowledge of CE, it was found that close to 70% of research participants indicated that they understood the meaning of the CE concept; however, only 12.9% of them had adequate knowledge to adopt and apply CE in their businesses and organisations [76]. Tam (2009) [26] argued that inadequate private sector education on waste management technology investment is a crucial barrier to adopting C&D waste-related technologies in the Australian industry relative to its Japanese equivalent.

A study in Australia [169] found that decisions on waste management practices in the building and construction sector are based on their financial return. The authors highlight the importance of imposing legislation to enhance waste management practices until such practices become culturally embedded in organisations across the supply chain. Therefore, it can be inferred that education and an associated cultural shift need to be aligned with encouragement and enforcement approaches. Only a few studies have examined the effect of education on the Australian C&D waste management system. The following section set out to identify educational opportunities that would enhance the industry's capacity to alter the present C&D waste management status quo and explore the impact of educational programs on C&D waste management in Australia.

11.2. Educational Opportunities in Australia

This section explores the extant educational opportunities for stakeholders involved in C&D waste supply chains. As illustrated in Figure 62, six common educational opportunities are provided by both the public and private sectors.

11.2.1. LinkedIn Learning

In recent years many universities and educational institutes across Australia have subscribed to this platform to unlock countless learning opportunities for their students and staff. LinkedIn Learning (LL), as a subsidiary of LinkedIn, is an online learning provider that offers video courses taught by industry experts in software, creative, and business skills. All the courses on LinkedIn fall into four categories: Business, Creative, Technology and Certifications and are designed for three levels: beginner, intermediate, and advanced. At the time of writing this article, thirty-one courses on this platform are directly or indirectly linked to C&D waste management.

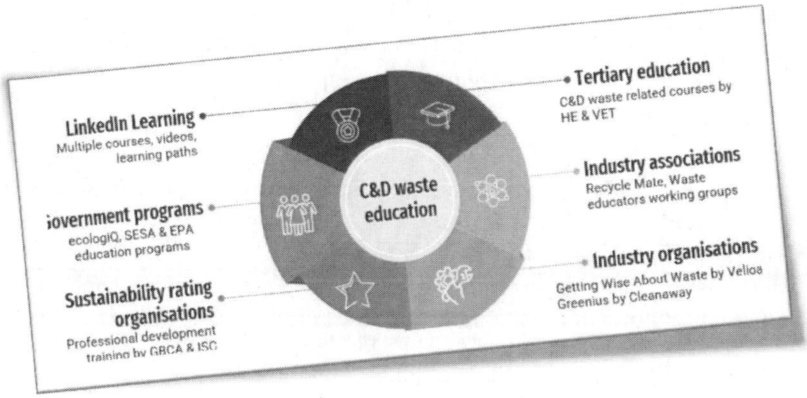

Figure 62. The most common educational opportunities in Australia.

11.2.2. Tertiary Education

The formal tertiary education system in Australia involves two primary levels: higher education (HE) and vocational education and training (VET, also known as TAFE: Technical and Further Education). HE focuses primarily on knowledge, theory, and critical thinking, and VET emphasises job-specific practical skills. The HE offers bachelor's degrees, graduate certificates, graduate diplomas, master's degrees, and doctoral degrees, and VET provides four levels of Certificates, diplomas and advanced diplomas. Currently, the Australian tertiary education system advances the industry's understanding of effective C&D waste management in three ways: (1) providing waste management-specific courses in VET, (2) providing core or elective courses on sustainable construction management, technologies, practices and planning in HE and (3) providing opportunities for research activities focusing on C&D waste management in HE. Table 27 shows the current VET courses specific to waste management and their description.

With increased awareness of construction environmental impacts, Australian universities started to incorporate topics related to waste management, planning and practices in their course contents. The waste content is primarily included in courses that deal with construction sustainability. Also, few courses were offered specific for waste management, including 'Sustainable Waste Management' that covers C&D waste management. Some of these courses are also provided individually through Open Universities Australia [170]. Research activities provided via research programs or courses at three levels (i.e., undergraduate, master's and PhD programs) represent a further potential for the education of people about the C&D waste stream within HE.

Table 27. VET courses on waste management offered in Australia

Course title	Duration	Course syllabus/description
Certificate II in Waste Management	1 year	Students learn how to stay safe at work and then correctly sort waste and recyclables according to environmental protection laws. Students can choose electives that will enable you to drive forklifts and waste collection vehicles
Certificate III in Waste Management	1 year	The course includes identifying and segregating waste, conducting waste resource recovery, identifying hazards and responding to waste management emergencies, following WHS and site procedures, operating waste processing plant, carrying out waste assessments, operating waste collection vehicles, administration and leadership skills
Certificate IV in Waste Management	1 year	It teaches essential aspects of operations and safety management at waste processing facilities. Students will learn how to build relationships with staff and clients while implementing waste management plans—in accordance with environmental protection laws

Students engaged in research-related courses at the undergraduate level may choose C&D waste management as their research topic, leading to Honours recognition from their host educational institutions. Students pursuing a Masters degree may undertake research on C&D waste management challenges via either the Masters by Research or Masters by Coursework programs. As the last point, there is a rising number of PhD candidates seeking to research C&D waste management for their doctoral theses. As shown in Figure 63, the number of thesis projects completed recently has seen a spike. According to Shooshtarian et al. (2021b) [171], this might be due to increasing government funding possibilities for investigating C&D waste management challenges.

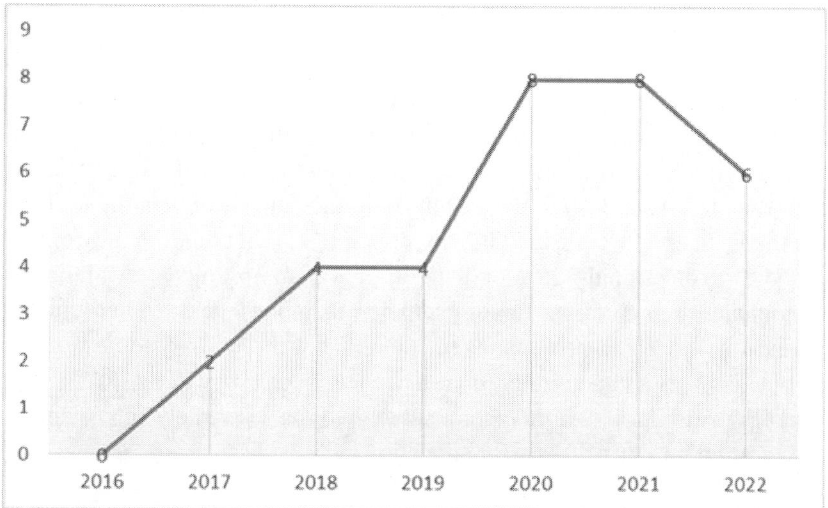

Source: Trove, National Library of Australia (2022).

Figure 63. Number of research theses related to C&D waste have recently completed in Australia.

11.2.3. Industry Associations

Industry associations play an important role in providing a collective voice for individual businesses within an industry. Additionally, they attempt to educate their members on current industry-specific issues and solutions. In Australia, several associations provide educational opportunities for stakeholders involved in C&D waste supply chains. Notably, the most active ones include the Institute of Engineers Australia (IEA), the Waste Management and Resource Recovery Association of Australia (WMRR), and the Australian Council of Recycling (ACOR). These associations frequently arrange waste conferences and exhibits to offer networking and educational opportunities for industry stakeholders. WMRR has facilitated state-based waste education working groups to strategise waste education initiatives at the state and national levels. These working groups consist of industry professionals, government representatives, and academics. ACOR designed a community-driven national recycling education program Recycle Mate, to increase public awareness of recycling issues and support recycling initiatives.

11.2.4. Sustainability Rating Organisations

These organisations primarily aim to improve sustainability in the Australian building and construction sector through their rating systems which recognize sustainability efforts in the building and infrastructure sectors [47]. Th. The main two organisations are the Green Building Council of Australia (GBCA) and the Infrastructure Sustainability Council (ISC). In addition, they organise periodic professional development training to promote awareness of sustainable construction among their members and other industry professionals. C&D waste management is one of the primary topics explored in these courses and activities. In a survey conducted by GBCA [172], it was found that 91% of its members considered CE as 'extremely important' and 'very important key of focus for education.

11.2.5. Industry Organisations

Various organisations operating in the waste recovery sector have launched educational programs that focus on behavioural change in the industry. Examples include Getting Wise About Waste and Greenius offered by Veolia and Cleanaway, respectively that are two major waste service providers in Australia. The former initiative is a fully interactive online learning portal for waste and recycling education– titled the EduPortal. The EduPortal provides clients with tailored information to help staff on-site across various positions, understand how to identify different waste streams and locate the correct receptacle within which to dispose of these materials.

11.2.6. Government Educational Programs

National and state governments have a pivotal role in providing and supporting educational programs for the building and construction sector. In most state waste strategy documents the role of education is highlighted [78]. Some examples of these programs include ecologiQ which is a Victorian Government program that was designed in 2019 to help Victorians incorporate construction materials into transport projects. The ecologiQ offers educational opportunities for industry stakeholders in order to maximise the usage of recycled materials in infrastructure projects. In addition, the federal government, through Skilling Educators for Sustainability Australia (SESA), offered an opportunity for trainers in the waste sector to increase their skills and capabilities in order to better help their colleagues in responding to waste concerns. Lastly, each state's EPA offers professional educational opportunities. These opportunities that might be obligatory aim to enhance the industry understanding of dealing with waste challenges.

11.3. Education's Influence on C&D Waste Management: Analysis of Evidence

In this section, the study analyses the impact of education on C&D waste educational programs. There is little evidence showing how the educational programs have led to increased sustainability in C&D waste management. However, it is documented that the quality of an educational programme may have a substantial influence on the attitudes and behaviours of industry professionals. For instance, a survey conducted by an Australian organisation [173] showed that waste educational programs would be more effective if they were primarily face-to-face, through networking and conferences and included onsite field studies. The research also attributed the success of waste education to providing accreditation or pathways into larger qualifications. The two main barriers to successful waste education included time and cost followed by not knowing a course is running and the lack of perceived value and relevance of an education program.

11.3.1. Models for Training Assessment

The Kirkpatrick model and the ADKAR Model have been identified as supporting foundations for change management and training assessments. The following sections describe those models in detail.

11.3.1.1. ADKAR Model

The ADKAR Model of Change Management [174] is an outcome-oriented change management method that aims to limit resistance to organizational change (Figure 64). Organizational changes often fail if employees don't fully understand the importance of getting on board the change or how to successfully make the change.

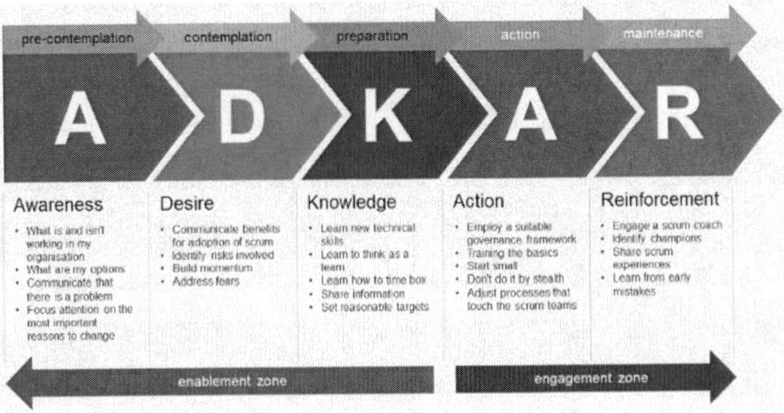

Source: Younes (2017) [174].

Figure 64. The components of the ADKAR model.

Employees often simply understand that a change is happening, and leaders are often not equipped to engage individuals effectively during the change and to manage any potential resistance [175].

The ADKAR Model addresses these challenges by equipping leaders with the right strategies and tools, and individuals with the right information, motivation and ability to successfully move through changes in the organization. By outlining the goals and outcomes of successful change, the ADKAR Model enables leaders and change management teams to focus their activities on what will drive individual change and therefore achieve organizational results.

11.3.1.2. Kirkpatrick Model [65]

The Kirkpatrick Model is a globally recognized method of evaluating the results of training and learning programs. It assesses both formal and informal training methods and rates them against four levels of criteria: reaction, learning, behaviour, and results (Figure 65). Kirkpatrick [65] asserts that training is evaluated using the four assessment levels described and that these are sufficient for holistic training evaluation. However, since its introduction, several other important evaluation models have been developed, many of which stem from the Kirkpatrick Model [65]. For example, the input-process-output (IPO) model [176] begins by identifying pre-training components (e.g., training materials, instructors, facilities) that impact efficacy as the input stage. The process stage focuses on the design and delivery of the training programs. Finally, the output stage essentially covers the same scope as the

Kirkpatrick Model. The six-stage evaluation model goes beyond assessment into training design and implementation [177].

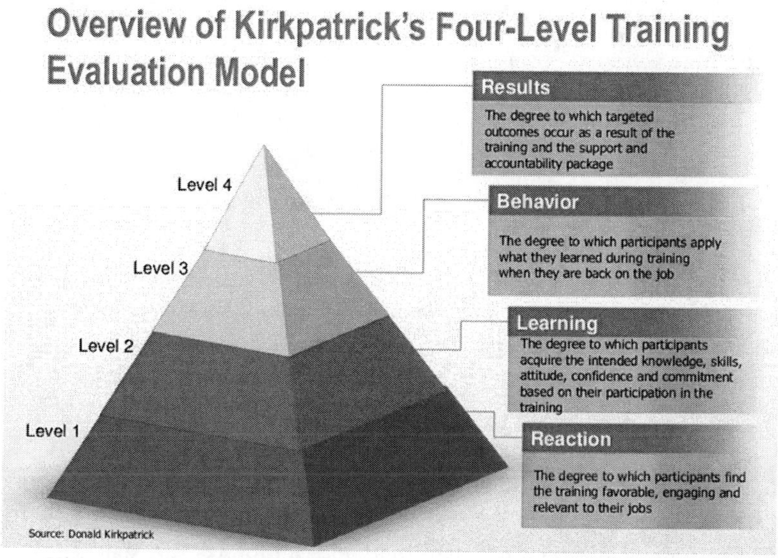

Source: Kirkpatrick (1959) [65].

Figure 65. Overview of Kirkpatrick's four-level training evaluation model.

The first stage identifies the goals of training, and the second stage assesses the design of a training program before implementation. The remaining four stages fall in line with Kirkpatrick's four levels. Kaufman and Keller (1994) [178] present a five-level evaluation model where Level 1 is expanded to include enabling, or the availability of resources, as well as a reaction; Levels 2 through 4 match the corresponding levels in the Kirkpatrick Model; Level 5 goes beyond the organization and presents a method of evaluating the training program on a societal level.

11.4. Conducting Pilot-Testing of a Training Module in Western Australia

In 2023, a training module on C&D waste management issues was developed in consultation with industry and government sectors. The module was pilot tested with one public infrastructure sector industry partner (Main Roads Western Australia) and a building public sector industry partner (Department of Communities WA). The content of the training packages was informed by the longitudinal study and included content related to unpacking the waste issue, circular economy principles, regulations, materials and market opportunities. This section presents the post-qualitative survey findings from two training sessions conducted on 9 March 2023. The detailed research approach is presented in Chapter 2.

11.4.1. Overview of Participants and Training Expectations

In the first session, overall, there were 21 participants including material engineers, project managers, and construction managers. A total of 12 participants (57%) responded to the post-training survey. One of the key objectives of this study was to determine if the training content met the participant's expectations. According to the survey results, ten participants said 'Yes' and two participants said No when they were asked if the training content met their expectations. One participant stated that

> "The training course is adaptable and relevant for MRWA staff."

Another participant described that the

> "Audience mostly aware of opportunities and the existing problem is the current blockages to these opportunities."

This participant further explained that a range of challenges included

> "Limits in specifications, supply, cost and absence of policy/legislation that enforce."

Survey participants described a range of expectations related to education and awareness, understanding the circular economy concept, material opportunities and case study examples. Table 28 presents a summary of the qualitative responses.

In the second session, there were 27 participants including project managers, construction managers and administrative staff. A total of 13 participants (48%) responded to the post-training survey. One of the key objectives of this study was to determine if the training content met the participant's expectations. According to the survey results twelve participants said Yes and one participant said No when they were asked if the training content met their expectations. Participants described a range of expectations related to education and awareness, understanding the circular economy concept, material opportunities and case study examples. Table 28 provides a summary of the qualitative responses. In addition to the above training expectations one participant suggested that the information was at a high level and I would have liked some detailed case studies with lessons learned.

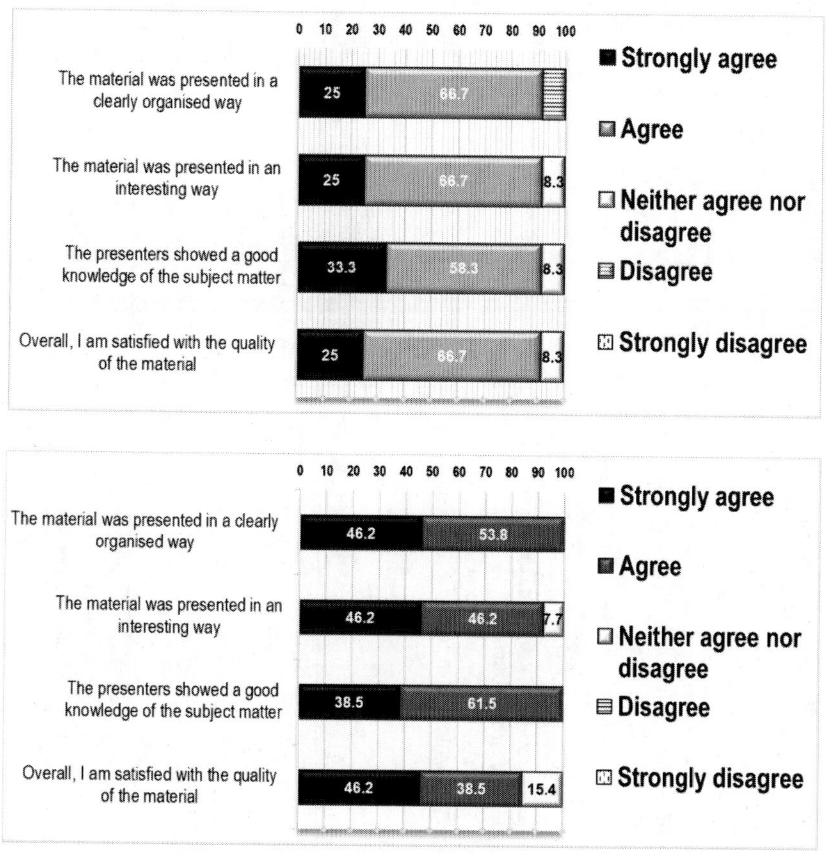

Figure 66. Participant responses on the quality of the training in session 1 (above) and session 2 (below)

Table 28. Summary of the training expectations in two sessions

Category (session 1)	Key quotes (session 2)	Category (session 2)	Key quotes (session 2)
Education and awareness	• "Improve my knowledge in the management of C&D materials and their reuse in infrastructure projects" • "Educational and informative session" • Raising and aligning awareness of sustainability	Understand C&D waste issues	• "Getting an overall view of construction and demolition issue and what is being done about it all" • "Gather more information on this topic" • "To gain knowledge"
Understand the circular economy concept	• "The training would be able to be more broadly adopted within MRWA to a larger group on the topic of circular economy" • "Introduction to circular economy" • "Whole of life sustainability considerations"	Learn about sustainable construction techniques	• "To understand how this training course can be applied to improving our sustainable construction techniques" • "Learn more about reusing waste" • "To be informed about how materials could be reused" • "To learn initiatives minimising waste in the building and construction sector" • "Understand solutions to mitigate site waste"
Identify material opportunities	• "Understand opportunities for reuse of materials" • "To recognise and improve design opportunities to reduce C&D waste" • "Provide an overview of previous projects" • "Understanding opportunities to pursue for recycling road construction material"	Strategies and identify opportunities for economic savings	• "I was hoping to find out more about the strategies that can be used, the dollar savings, and detailed examples where this had occurred." • "Understand how could building material be used in a sustainable manner while reducing unusable waste"
Learn about case study examples	• "Case studies for project managers in planning on how projects have can have best outcomes and set projects up for success in practical ways"		

11.4.2. Quality of the Training

As shown in Figure 66, nearly 92% (n = 11) of the participants in the first session either strongly agreed or agreed with the statement that the material was presented in a clearly organised way. One participant disagreed with this statement. A total of 11 participants strongly agreed (25%) or agreed (67%) that the material was presented in an interesting way. One participant neither agreed nor disagreed (8%) with this statement. Out of 12 participants, 11 participants strongly agreed (33%) or agreed (58%) that the presenters showed a good knowledge of the subject matter. One participant neither agreed nor disagreed (8%) with this statement.

A total of 11 participants strongly agreed (25%) or agreed (67%) that they were satisfied with the quality of the material. One participant neither agreed nor disagreed (8%) with this statement. All participants agreed that the mix between PowerPoint slides and activities was suitable.

In the second session, all 13 participants either strongly agreed (46%) or agreed (54%) to the statement that the material was presented in a clearly organised way (Figure 66). A total of 12 participants either strongly agreed (46%) or agreed (46%) that the material was presented in an interesting way. One participant neither agreed nor disagreed (8%) with this statement. All 13 participants either strongly agreed (39 %) or agreed (62%) that the presenters showed a good knowledge of the subject matter. Out of the 13 participants, 11 participants either strongly agreed (46%) or agreed (39%) that they were satisfied with the quality of the material. Two participants neither agreed nor disagreed (15%) with this statement. Lastly, all 13 participants agreed that the mix between PowerPoint slides and activities was suitable. For example, one participant said,

> "I liked the break with the little interactions to break up the course. That was refreshing than just listening to talking heads" and another participant said that "content and pace was perfect."

11.4.3. Key Learning and Opportunities for Improvement

In the first session, out of 12 survey participants, 10 stated they learned something new. Table 29 presents a summary of the responses to the new learnings of the participants in the first session under the categories of reusing and recycling opportunities, state/organizational level requirements, case study examples and holistic approach to circular economy. Out of the total participants, ten participants said that the course is easy to apply to their work areas, while two participants said No. For example, one participant described that:

> "The next steps in the role are to address the barriers to the opportunities, not to identify the opportunities or problem statement."

The participants shared detailed feedback on key learnings of the training course. One participant specifically stated that this

> "The training course is adaptable and relevant for MRWA staff".

Table 30 presents a summary of the qualitative responses on their learnings under the key categories including waste hierarchy and relevant opportunities; circular economy opportunities; and learning from global best practices.

Several participants shared insights on how the training course can be improved. One participant highlighted that there could be more discussion on whole-of-life/planning/development considerations, and case study analysis. One participant suggested to share:

> "Some more examples of reused and recycled materials in industry that are currently being undertaken". Another participant pointed out that it is important to "tailor to audience and problems that are resulting in the lack of uptake of opportunities."

In the second session, all 13 participants stated that they learned something new. Out of the total, 12 participants said that the course is easy to apply in their work areas. While one participant said no as the role is more focused on administration aspects. The participants shared detailed feedback on key learnings of the training course. Table 30 presents a summary of the qualitative responses on their learnings under the key categories of sustainable development goals, waste hierarchy and relevant opportunities, significance of the environmental impacts and circular economy opportunities.

Table 29. Summary of the new learnings in two sessions

Category (session 1)	Key quotes (session 2)	Category (session 2)	Key quotes (session 2)
Reusing and recycling opportunities	• "Importance for recycling concrete"	Recycling techniques and opportunities	• "The way waste from demolitions is being recycled" • "Using recyclable for new projects" • "Different ways to look at materials and waste"
State/ organizational level requirements	• "Recycling and reuse are of high priority and supported by Infrastructure Sustainability" • "Different organisation/ state sustainability requirements"	Environmental impacts	• "Carbon footprint of building materials"
Case study examples	• "Interstate project examples" • "Other projects using plastic noise walls"	Circular economy practices	• "There is an effort being made to reduce construction waste through a circular economy" • "Practical ways of doing so, the constraints we may encounter, what variables we need to account for" • "Examples of circular economy"
Holistic approach to the circular economy	• "A more holistic project approach to implementing the circular economy"	Industry challenges	• "The complexity of knowing what we know versus how it can be implemented" • "Understanding more about the constraints facing the industry and new innovative solutions entering the market"

Table 30. Summary of the key learnings in two sessions

Category (session 1)	Key quotes (session 1)	Category (session 2)	Key quotes (session 2)
Waste hierarchy and relevant opportunities	• "Extents of reuse opportunities" • "Consideration to the reuse/recycle of C&D materials very early in project development and how to manage the materials"	Sustainable development goals	• "The broad sustainable development goals which then need to trickle down to our agency's goals" • "This space is evolving and expanding and we at communities can apply this teaching in our directorate"
Circular economy opportunities	• "Importance of extending the utility phase of a material to reduce stress on virgin materials" • "Circular economy and potential construction application for MRWA major projects" • "Road construction has a significant potential for circular economy contributions"	Significance of the environmental impacts	• "The high number of emissions the construction industry is responsible for"
Learning from local and global best practices	• "State approaches" • "There is a lot of work still to do to get Australia up to leading country standards. Our thinking is where Germany, Netherlands and Sweden were decades ago"	Waste hierarchy and relevant opportunities	• "This space is evolving and expanding and we at communities can apply this teaching in our directorate" • "In our current work, we can get our builders/partners to be more conscious" • "Where and how we can reuse waste materials"
		Circular economy opportunities	• "Check-list example is a tangible tool for projects, this idea should be further explored" • "Use of materials in a circulatory sense" • "Consider how our housing designs and specifications might impact the circular economy" • "The design stage is critical"

Table 31. Summary of strategies for successful design and implementation of waste educational programs

Step	Description of strategies	
1	• Engage external evaluator to conduct assessment activities • Create a forum to generate discussion and promote interaction among stakeholders	• Build organisational capacity and stakeholders' support to collaborate to evaluate program success measures
2	• Create a logic model to assist stakeholders understand the program's objectives and how it is accomplishing them	• Develop a logic model that shows your program's logic and the links between inputs (resources used), outputs (activities and audience), and benefits (outcomes and impacts in the short, intermediate and long-term)
3	• Identify why the evaluation is planned and who will use the evaluation's results which determine the method of evaluation • Develop indicators that help demonstrate whether program goals have been achieved	• Set evaluation questions based on the logic model components (created in Step 2), which also consider the concerns of multiple stakeholders
4	• Identify what type of data to collect and how to collect that data by reviewing the most effective methods	
5	• Devising a work plan with deadlines for all key aspects of the evaluation to keep the process manageable	• Identify expectations for staff and evaluation team members • Pilot-test instruments to ensure that directions are clear and that measures are appropriate for the target audience
6	• Ensure to analyse data at the appropriate level (e.g., intermediate or advanced)	
7	To determine how results are shared based on target audience and program goals	
8	• Communicate the program's successful results to demonstrate the impact it has on your target audience or the community. • Address organisation and planning problems to make the implementation of program more successful and increase participant satisfaction • Change content based on participants' feedback or other evaluation results	• Use what is learned from the evaluation to identify ways to increase awareness of your program's offerings • Consider investing in certain components of the program if some components provide less than ideal results than others • Learn from the mistakes and limitations of past evaluations to avoid these pitfalls in the future

Several participants shared insights on how the training course can be improved. One participant stated that

> "Presentation should be circulated with broader Industry, particularly key project home building partners - they provide pricing and designs against their standardised material selections and pressure should be placed on including more recycled products. The consideration of cost implications for industry and improvements to current legislation will drive change".

Another participant added to this by stating *"a circular economy example would be helpful i.e., carbon reduction, material waste reduction, reuse of x materials, $ savings etc"*. One participant suggested immediate action by stating

> "Although a quick win would be suppliers taking back our surplus flooring or it being sent to our maintenance contractor as this is an issue as the surplus product is removed from our new builds before tenancy."

11.5. Quality Assessment of Educational Programs

As stated before, waste educational programs aim to influence attitudes and behaviours in the building and construction sector and the waste recovery industry. To do this, educational programmes must be revamped, and authorities must concentrate on maximising effect. This section offers strategies for achieving education providers' goals. These solutions rely on a framework developed by Michaela Zint of the University of Michigan [179], as shown in Figure 67.

This framework guides improvements in educational programs in four phases (understanding the program, planning evaluation, implementing evaluation and improving the program) and through eight steps. These eight steps are described in Table 31.

In addition to evaluation, there are other strategies that may help program educational program achieve their goals. These strategies are provided below:

- Provide accreditation or pathways into larger qualifications
- Devise mandatory education for individuals engaged in C&D waste generation and handling
- Alignment between various waste educational initiatives

- Financial assistance from the government for participation in educational programmes
- Building capacity in organisations to enable individuals' participation in waste education

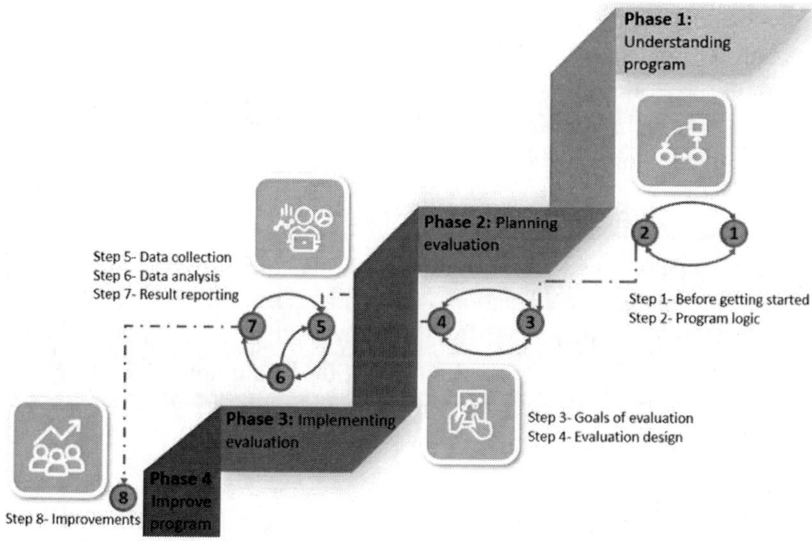

Source: Adopted from Meera (2022) [179].

Figure 67. A framework to evaluate an educational program effectiveness.

Conclusion

Education, enforcement, and encouragement are the three driving forces that enable the design and implementation of a successful C&D waste management system. In the Australian context, the role of education in advancing C&D waste is yet to be fully uncovered. This chapter attempts to capture the educational opportunities that will facilitate behavioural and attitudinal changes towards better waste management in the building and construction sector. The contextual literature review showed that in the Australian context, there are six major channels for waste education.

Two training packages were co-created targeting industry practitioners involved in management and decision making for solid waste systems which recycle and dispose of waste generated as a result of construction, deconstruction, renovation and demolition of buildings and structures.

Specifically, the intended audience included design and structural engineers, project managers and construction managers.

The program was pilot-tested with Western Australia based infrastructure and building-focused organisations. The training materials included content related to unpacking the waste issue, circular economy principles, regulations, materials and market opportunities. The training effectiveness was evaluated through a survey inspired by the Kirkpatrick model [65]. A key practical contribution from this study includes targeted opportunities for capacity building through multiple industry-driven strategies that enable the construction sector to achieve circular economy objectives for extended use of resources.

The present chapter contributes to the theory and practice of waste education in the Australian building and construction sector by providing suggestions that enable educational programs to achieve their intended goals. This chapter recommends performing comparative studies to examine the efficacy of relevant educational initiatives in developing a CE in the sector and reducing C&D waste disposal in Australia.

Chapter 12

Conclusion

12.1. Introduction

With the continuous increase in C&D waste recovery rates in Australia, the next challenge is to establish a system for the sustainable application of PwRC in construction projects. Therefore, new initiatives aiming to address C&D waste resources should focus on this issue. This book, spanning eleven chapters, delves into significant initiatives crucial for governmental bodies and industries in attaining this objective. These initiatives include comprehending the drivers affecting the adoption of PwRC in construction, evaluation of recycled product certification schemes, employing circular design approaches, and fostering educational opportunities. Collectively, the book aims to boost business prospects, provide advantages to product manufacturers, and establish sustainable circular supply chains. Subsequent sections outline the key highlights from the preceding chapters.

12.2. The Use of PwRC in Construction Projects in the Global Context

It is important to understand the main barriers, motivations and opportunities for increasing the utilisation of PwRC in the building and construction sector. *In Chapter 1,* first, these factors were determined through analysis of the existing literature. The analyses showed that the main enablers of using PwRC in the construction projects include (1) community awareness and education on PwRC, (2) supportive regulations, policies and specifications, (3) enabling sustainability programs, (4) recycled product certification and (5) targeted technologies and innovative practice. These enablers can mitigate the impacts of some of these barriers (1) cost and energy, (2) lack of knowledge, (3) unsupportive regulations, (4) negative perceptions, (5) PwRC's poor performance, (6) limited technology, and (7) lack of market.

Second, in *Chapters 3-7,* these factors were investigated through a multiple case study approach in the Australian context. The top four barriers to using PwRC in construction projects are 'unsupportive regulations,' 'limited availability of PwRC,' 'lack of expertise and understanding of PwRC utilisation' and 'inconsistency of PwRC quality and performance and issues around the warranty.' The research findings will have policy implications as the 'unsupportive regulations' were recognised as the top barrier. The current climate of the waste recovery system in Australia, and possibly all around the world, means governments are being urged to devise policies that facilitate the utilisation of PwRC and drive a CE in the building and construction sector. As such, policymakers can use the information provided in this study to develop policies that provide a level playing field for all parties operating in PwRC supply chains. For instance, informed policies can incentivise the production and purchasing of PwRC in the sector.

According to the responses from the research participants, the barriers identified in this study are to be primarily addressed through 'education, investigation, and demonstration activities' and 'effective project management planning.' Further research should investigate how the outlined strategies may change stakeholders' attitudes and result in the optimised use of PwRC in construction. In addition, the policy reform initiatives must pay particular attention to these measures as part of a road map leading to the optimised use of PwRC.

Additionally, the two main motivations for using these resources were found to be 'environmental benefits' and 'ensuring competitive advantage and futureproofing.' Accordingly, new policies can give the motivating factors found in this study a greater weight in the equation of PwRC utilisation. The research also has a practical contribution to the industry through a suite of actionable strategies to increase the uptake of PwRC. The two major motivations, referring to two of the three pillars of sustainability (economy and environment), provide a basis for organisational changes to ensure the achievement of sustainability in construction activities.

The thematic and descriptive analyses revealed that the study factors might affect the utilisation of PwRC to a different degree and/or in different directions. Hence, while the aggregated data in this study provide useful insight into PwRC application, precise planning for using these resources is subject to a wide array of contextual variables, including evident and latent constraints and opportunities.

12.3. Application of Recycled Product Certification

In *Chapter 8,* the use of recycled product certification was thoroughly investigated in three steps. First, the literature review was used to identify the main enablers and drivers of using RPC in construction projects. The advantages included (1) cost associated with their application (2) lack of cost justification (3) self-certification as an alternative, (4) lack of alignment between RPC and sustainability rating systems, (5) unreliable certifiers and (6) lack of collaboration between parties involved in certifying PwRC. The main enablers, according to the existing literature, are (1) robust legislation (2) promotional activities (3) education of stakeholders, (4) contractual obligation (5) training qualified material engineers, and (6) simple set up.

In the second step, the understanding of Australian stakeholders operating in the built environment sector concerning RPC application was explored using a case study approach. The findings offered useful information about the characteristics of RPC with a focus on the Australian context and generalisability to similar contextual conditions by analysing four case studies in which PwRC were utilised.

Analysis of four case studies showed that only one out of four case projects applied this scheme. The study also listed the top seven limitations of RPC application, namely unjustified application in the presence of product specifications, the abundance of RPCs with different requirements, lack of transparency about their requirements, limited applicability for certain products, unjustified cost, certifiers' poor performance and, recyclers' limited capacity to be certified. To address these limitations, this chapter proposed several strategies aimed at various key stakeholders (government, end-users, industry associations, waste recyclers, certifiers, education providers and sustainability rating organisations) to make the RPC application business-as-usual, when proven useful.

The building and construction sector can use the research results to refine their working business models and practices to optimise the uptake of PwRC in construction projects. Lastly, policymakers may leverage the findings in this research to develop informed policies that provide a level playing field for all those operating in the PwRC supply chain.

12.4. Circular Design Principles

Chapter 9 set out to develop CE guidelines for the C&D waste management of the Australian built environment sector. Although various design methods exist to minimise C&D waste, DoW guidelines have yet to be clearly outlined. This chapter sets a framework for the successful implementation of these guidelines. In doing so, a descriptive literature review was conducted to understand the current global and Australian CE practices in the sector. This study also identified the main clusters of barriers and enablers of CE implementation in the built environment sector. Analysing these factors will help decision-makers make informed decisions about the extent to which CE principles can be applied in the sector. Finally, this paper outlined industry-specific guidelines for implementing CE practices. Below are the main takeaways from these guidelines that will have sustainability management implications in the built environment sector:

- The guidelines present 16 strategies with detailed actions that apply to all stages in the value chain, from design, manufacture, and operation, to EoL. The project managers in the sector can adapt and apply these strategies where possible.
- The strategies encourage the sector to move away from the traditional, linear way of managing the built environment by keeping construction materials in the loop and improving waste management performance and using the built environment to regenerate nature (e.g., climate-resilient design). The latter is particularly important in delivering future projects where the demonstration of achieving (carbon) emission reductions through climate-resilient design is necessary.
- The seven categories identified as having a literature-specific focus (CE/Zero waste, Construction Innovation, DfD/Circular design, Digital Technology, Project/Supply Optimisation, Resource Management, Reuse, and Sustainability Assessment) served as a foundation to develop the three main scope areas of the proposed guidelines (circularity practices, resource management, and innovation and optimisation). Paying attention to these scope areas is necessary from the project management perspective to deal with barriers they face throughout the process of transforming to circularity.

- The circularity practices have a strong focus on various design concepts, namely, DfD, design for flexibility and adaptability, design for long life, design out hazardous/pollutant materials, and climate-resilient design. Resource management strategies are relevant when the material reaches its EoL. Innovation and optimisation strategies look to make the most of technological breakthroughs to realise circular built environments. Hence, designers and project managers in the sector need continuous education and training on the latest technological advancements to ensure that circularity will be business as usual.
- The literature review uncovered targeted strategies and actions that contribute to the DoW and pollution and keeping resources in use (two of the EMF CE principles)

Most notably, the CE guidelines for the built environment suggested in this study should be seen as a dynamic framework that should be updated and revised with new research, good practices, technologies, and continuous learning.

12.5. Circular Design Case Study

Chapter 10 focused on the Legacy Living Lab (L3) project by Curtin University in Perth, Australia—a pioneering example of DfD and the circular economy in action. The project aimed for a versatile, adaptable space that can be efficiently deconstructed and reassembled.

The L3 encountered various challenges, including material sourcing, joint fixing, and the absence of design standards tailored to Australian contexts [14, 19]. Despite prioritizing CE principles, L3 did not achieve full circularity. A significant takeaway is the imperative for cross-sectoral collaboration. The project highlighted that effective DfD implementation is a collaborative venture involving design architects, developers, contractors, and policymakers. The modular approach has wide-ranging applicability, especially in temporary housing and disaster relief, underscoring the flexibility inherent in DfD principles.

The challenges predominantly stemmed from regulatory barriers, including intricate building codes. Conversely, L3 opened avenues for academic inquiry, community engagement, and commercial alliances.

L3 serves as a pioneering example of construction aligned with circular economy principles. The substantial environmental benefits present compelling reasons for adopting DfD in future endeavours. As a living lab, L3's economic feasibility was not assessed, constituting a major limitation. Nevertheless, the project illuminated critical technical and constructability aspects of DfD within a local setting.

Given L3's positive impact and insights gained, there exists a compelling case for the wider integration of DfD in multiple sectors. Future research should aim to alleviate regulatory barriers and promote industry-wide standards.

12.6. The Impact of Education on the Use of PwRC

Education, enforcement, and encouragement are the three driving forces that enable the design and implementation of a successful C&D waste management system. In the Australian context, the role of education in advancing C&D waste is yet to be fully uncovered. *Chapter 11* attempts to capture the educational opportunities that will facilitate behavioural and attitudinal changes towards better waste management in the building and construction sector. The contextual literature review showed that in the Australian context, there are six major channels for waste education.

Two training packages were co-created targeting industry practitioners involved in management and decision making for solid waste systems which recycle and dispose of waste generated as a result of construction, deconstruction, renovation and demolition of buildings and structures. Specifically, the intended audience included design and structural engineers, project managers and construction managers. The program was pilot tested with Western Australia based infrastructure and building focused organisations. The training materials included content related to unpacking the waste issue, circular economy principles, regulations, materials and market opportunities. The training effectiveness was evaluated through a survey inspired by the Kirkpatrick model [65]. A key practical contribution from this study includes targeted opportunities for capacity building through multiple industry-driven strategies that enable the construction sector to achieve circular economy objectives for extended use of resources.

References

[1] Zheng, L., Wu, H., Zhang, H., Duan, H., Wang, J., Jiang, W., Dong, B., Liu, G., Zuo, J. and Song, Q. 2017. Characterizing the generation and flows of construction and demolition waste in China. *Construction and Building Materials*, 136, pp.405-413. https://doi.org/10.1016/j.conbuildmat.2017.01.055.

[2] Guzmán, J., Marrero, M., Montes-Delgado, M.V. and Ramirez-de-Arellano, A. 2009. A Spanish model for quantification and management of construction waste, *Waste Management*, 29(9), pp. 2542-2548. https://doi.org/10.1016/j.wasman.2009.05.009.

[3] Menegaki, M. and Damigos, D. 2018. A review on current situation and challenges of construction and demolition waste management, *Current Opinion in Green and Sustainable Chemistry*, 13, pp. 8-15. https://doi.org/10.1016/j.cogsc.2018.02.010.

[4] Deng, D. 2022. An insight into China's construction and demolition waste recycling industry. Invest Northern Ireland. https://bit.ly/3ktLNi1.

[5] US EPA. 2019. *Advancing sustainable materials management: 2017 Fact sheet.* EPA 530-F-19-007. Environmental Protection Agency. https://bit.ly/3OhD3WR.

[6] Allied Analytics LLP. 2023. *Construction and demolition waste recycling market is booming worldwide to reach $149.2 billion by 2027.* Newswires. https://bit.ly/3IUtWsO.

[7] IMARC Group. 2023. *Construction and demolition waste management market: global industry trends, share, size, growth, opportunity and forecast 2023-2028.* SR112023A1230. The International Market Analysis Research and Consulting Group. https://bit.ly/3EXA1n0

[8] Report Ocean. 2021. *Construction and demolition waste recycling market research report.* Report Ocean. Chicago, USA.

[9] Blue Environment. 2022. *National Waste Report 2022*. The Department of Climate Change, Energy, the Environment and Water. Canberra, Australia. https://shorturl.at/bhvKO.

[10] ABRELPE 2022. *Overview of solid waste in Brazil in 2022.* Brazil: Brazilian Association of Public Cleaning and Special Waste Companies.

[11] Destatis. 2021. *Brief overview waste balance*. Waste management. Federal Statistical Office of Germany. https://shorturl.at/fDFP8.

[12] BMTPC. 2021. *Annual report: 2020-2021*. New Delhi, India: Ministry of Housing and Urban Affairs: Building Materials & Technology Promotion Council. https://shorturl.at/aF157.

[13] Ministry of Land Infrastructure Transport and Tourism. 2020. *Results from the survey on construction and demolition waste in 2018*. Japan.

[14] UK Government Statistical Service 2021. *UK statistics on waste*. https://shorturl.at/hkDMP.

[15] US EPA 2020. *Assessing trends in materials generation and management in the United States. Advancing Sustainable Materials Management: 2018 Fact Sheet.* US EPA. https://shorturl.at/dLOP4.

[16] The Federal Service for State Statistics. 2019. Moscow, Russia. https://eng.rosstat.gov.ru/folder/160366.

[17] ISPRA. 2020. *Italian environment: Trend and legislation*. Roma, Italy: Division for Environmental Information, Statistics and Reporting: Italian Institute for Environmental Protection and Research. https://rb.gy/n1u04f.

[18] DEME. 2021. *Treatment performance of the construction products and materials in the building sector*. Paris, France: The French Agency for Ecological Transition.

[19] Li, J., Yao, Y., Zuo, J. and Li, J. 2020. Key policies to the development of construction and demolition waste recycling industry in China. *Waste Management*, 108, pp. 137-143. https://doi.org/10.1016/j.wasman.2020.04.016.

[20] Active Sustainability. 2020. *Expanding reuse opportunities for recycled construction materials*. Perth, Australia. https://rb.gy/p7ilyt.

[21] Park, J. and Tucker, R. 2017. Overcoming barriers to the reuse of construction waste material in Australia: A review of the literature. *International Journal of Construction Management*. 17, pp.228-237. https://doi.org/10.1080/15623599.2016.1192248.

[22] Shooshtarian, S., Caldera, S., Maqsood, T. and Ryley, T. 2020. Using recycled construction and demolition waste products: A review of stakeholders' perceptions, decisions, and motivations. *Recycling*, 5, pp.31. https://doi.org/10.3390/recycling5040031.

[23] di Maria, A., Eyckmans, J. and Van Acker, K. 2018. Downcycling versus recycling of construction and demolition waste: Combining LCA and LCC to support sustainable policy making. *Waste Management*, 75, pp. 3-21. https://doi.org/10.1016/j.wasman.2018.01.028.

[24] Umar, U. A., Shafiq, N., Malakahmad, A., Nuruddin, M. F. and Khamidi, M. F. 2017. A review on adoption of novel techniques in construction waste management and policy. *Journal of Material Cycles and Waste Management*. 19, pp. 1361-1373. https://doi.org/10.1007/s10163-016-0534-8.

[25] Oyedele, L. O., Ajayi, S. O. and Kadiri, K. O. 2014. Use of recycled products in UK construction industry: An empirical investigation into critical impediments and strategies for improvement. *Resources, Conservation and Recycling*. 93, pp. 23-31. https://doi.org/10.1016/j.resconrec.2014.09.011.

[26] Tam, V. W. 2009. Comparing the implementation of concrete recycling in the Australian and Japanese construction industries. *Journal of Cleaner Production*. 17, pp. 688-702. https://doi.org/10.1016/j.jclepro.2008.11.015.

[27] Colin, R. 2019. *Systems for Reuse, Repurposing and upcycling of existing building components*. Ph.D. Thesis, University College London, London, UK. https://rb.gy/qylu0o.

[28] Dunant, C. F., Drewniok, M. P., Sansom, M., Corbey, S., Allwood, J. M. and Cullen, J. M. 2017. Real and perceived barriers to steel reuse across the UK construction value chain. *Resources, Conservation and Recycling*. 126, pp.118-131. https://doi.org/10.1016/j.resconrec.2017.07.036.

[29] Zou, P., Hardy, R. and Yang, J. 2015. *Barriers to building and construction waste reduction, reuse and recycling: A case study of the Australian capital region.* In Proceedings of the Building Today—Saving Tomorrow: Sustainability in Construction and Deconstruction Conference, Auckland, New Zealand, 15–17 July. https://rb.gy/ez5hl3.

[30] Hiete, M., Stengel, J., Ludwig, J. and Schultmann, F. 2011. Matching construction and demolition waste supply to recycling demand: A regional management chain model. *Building and Research Information*. 39, pp.333-351. https://doi.org/10.1080/09613218.2011.576849.

[31] Wang, J., Li, Z. and Tam, V.W.Y. 2014. Critical factors in effective construction waste minimization at the design stage: A Shenzhen case study, China. *Resources, Conservation and Recycling*. 82, pp. 1-7. https://doi.org/10.1016/j.resconrec.2013.11.003.

[32] Begum, R. A., Satari, S. K. and Pereira, J. J. 2010. Waste generation and recycling: Comparison of conventional and industrialized building systems. *American Journal of Environmental Sciences*. 6, pp. 383-388. https://doi.org/10.3844/ajessp.2010.383.388.

[33] Denyer, D. and Tranfield, D. 2009. Producing a systematic review. In *The Sage Handbook of Organizational Research Methods*, Buchanan, D. A. and Bryman, A. (Eds.). Sage Publications, Thousand Oaks, CA, USA. https://rb.gy/pkbfkm.

[34] Sustainability Victoria. 2015. *Market Summary—recycled brick, stone and concrete*. Sustainability Victoria: Melbourne, Australia. https://rb.gy/303k82.

[35] Liu, J., Nie, J. and Yuan, H. 2020. Interactive decisions of the waste producer and the recycler in construction waste recycling. *Journal of Cleaner Production*. 256, 120403. https://doi.org/10.1016/j.jclepro.2020.120403.

[36] Edge Environment Pty. 2012. *Construction and demolition waste guide—recycling and re-use across the supply chain*. Sydney, Australia. https://rb.gy/8t8vyk.

[37] Bolden, J., Abu-Lebdeh, T. and Fini, E. 2013. Utilization of recycled and waste materials in various construction applications. *American Journal of Environmental Sciences*. 9, pp.14-24. https://doi.org/10.3844/ajessp.2013.14.24.

[38] Ghaffar, S. H., Burman, M. and Braimah, N. 2020. Pathways to circular construction: An integrated management of construction and demolition waste for resource recovery. *Journal of Cleaner Production*. 244, 118710. https://doi.org/10.1016/j.jclepro.2019.118710.

[39] He, L. and Yuan, H. 2020. Investigation of construction waste recycling decisions by considering consumers' quality perceptions. *Journal of Cleaner Production*. 259, 120928. https://doi.org/10.1016/j.jclepro.2020.120928.

[40] Sánchez, I. G. and Lauritzen, E. K. 2004. *Integrated decontamination and rehabilitation of buildings, structures and materials in urban renewal-a European project for a sustainable city concept*. In Proceedings of the International RILEM

Conference on the Use of Recycled Materials in Buildings and Structures, Barcelona, Spain, 8–11 November 2004.

[41] Spiegel, R. and Meadows, D. 2010. *Green building materials: a guide to product selection and specification*. John Wiley & Sons, Hoboken, NJ, USA.

[42] Henry, M. and Kato, Y. 2012. Perspectives on sustainable practice and materials in the Japanese concrete industry. *Journal of Materials in Civil Engineering*. 24, pp. 275-288. https://doi.org/10.1061/(ASCE)MT.1943-5533.0000388.

[43] Knoeri, C., Binder, C. R. and Althaus, H. 2011. Decisions on recycling: Construction stakeholders' decisions regarding recycled mineral construction materials. *Resources, Conservation and Recycling*. 55, pp. 10391050. https://doi.org/10.1016/j.resconrec.2011.05.018.

[44] Mak, T. M. W., Yu, I. K. M., Wang, L., Hsu, S., Tsang, D. C. W., Li, C. N., Yeung, T. L. Y., Zhang, R. and Poon, C. S. 2019. Extended theory of planned behaviour for promoting construction waste recycling in Hong Kong. *Waste Management*. 83, pp. 161-170. https://doi.org/10.1016/j.wasman.2018.11.016.

[45] Huang, Y., Bird, R. G. and Heidrich, O. 2007. A review of the use of recycled solid waste materials in asphalt pavements. *Resources, Conservation and Recycling*, 52, pp. 58-73. https://doi.org/10.1016/j.resconrec.2007.02.002.

[46] Eisenstein, W., Fuertes, G., Kaam, S., Seigel, K., Arens, E. and Mozingo, L. 2017. Climate Co-Benefits of Green Building Standards: Water, Waste and Transportation. *Building Research and Information*. 45, pp. 828-844. https://doi.org/10.1080/09613218.2016.1204519.

[47] Shooshtarian, S., Maqsood, T., Wong, P., Khalfan, M. and Yang, R. 2019. *Green construction and construction and demolition waste management in Australia*. In Proceedings of the 43rd AUBEA Conference: Built to Thrive: Creating Buildings and Cities That Support Individual Well-Being and Community Prosperity, Noosa, Australia, 6–8 November 2019. https://rb.gy/zxyqm3.

[48] Lu, W., Chi, B., Bao, Z. and Zetkulic, A. 2019. Evaluating the effects of green building on construction waste management: A comparative study of three green building rating systems. *Building and Environment*. 155, pp. 247-256. https://doi.org/10.1016/j.buildenv.2019.03.050.

[49] Leipold, S. and Petit-Boix, A. 2018. The circular economy and the bio-based sector—Perspectives of European and German stakeholders. *Journal of Cleaner Production*. pp.1125-1137. https://doi.org/10.1016/j.jclepro.2018.08.019.

[50] Moher, D., Liberati, A., Tetzlaff, J. and Altman, D.G. 2009. Prisma Group. Preferred reporting items for systematic reviews and meta-analyses: The PRISMA statement. *PLOS Medicine*. 6, e1000097. https://doi.org/10.1371/journal.pmed.1000097.

[51] Curtin University. 2021. *Legacy Living Lab*. Curtin University, Perth, Australia. Available at https://l3.curtin.edu.au/

[52] Shooshtarian S., Maqsood T., Wong P. S. P., Caldera S., Ryley T., Zaman A. and Caceres Ruiz AAM. 2023. Circular economy in action: The application of products with recycled content in construction projects: A case study approach. *Smart and Sustainable Built Environment*. https://doi.org/10.1108/SASBE-08-2023-0213.

[53] Armao, J. 2021. *Burwood Brickworks recognised as the world's most sustainable shopping centre*, Available at: https://shorturl.at/aowNQ.

[54] MainRoads Western Australia. 2022. *'Tonkin gap project and associated works' MainRoads Western Australia*, accessed 13 May 2023. https://shorturl.at/xFHIY.

[55] TABEC. 2022. *115 Hamilton Hill Stage 1*, TABEC Civil Engineering Consultants, Perth, Australia. https://rb.gy/jhffsl.

[56] Victoria's Big Build. 2021. *Mordialloc Freeway*, Major Road Projects Victoria, Melbourne, Australia. https://rb.gy/l58234.

[57] International Living Future Institute. 2023. *Living building challenge.* https://living-future.org/lbc/.

[58] Green Building Council of Australia. 2023. *Rating system.* https://bit.ly/3HhC3jO.

[59] Infrastructure Sustainability Council. 2022. *IS ratings.* https://bit.ly/3wx1Z4N.

[60] Urban Development Institute of Australia. 2022. *National awards.* https://bit.ly/3CUwwgg.

[61] VicRoads. 2022. *Standard sections - 400 series - Asphalt and surface treatments 407 dense graded asphalt.* https://bit.ly/3XqErtW.

[62] Government of Western Australia. 2021. *Roads to Reuse product specifications: Recycled road base and recycled drainage rock.* https://bit.ly/3D0tibd.

[63] IPWEA. 2016. *IPWEA/WALGA specification for the supply of recycled road base.* https://bit.ly/3wj1tqH.

[64] Global GreenTag. 2023. T*he world's best eco products.* Certified. https://bit.ly/3JmhGmo.

[65] Kirkpatrick, D. L. 1959. Techniques for evaluating training programs. *Journal of the American Society of Training Directors*, 13, 3–9. https://shorturl.at/wxUV6.

[66] Victoria State Government. 2022. *Recycling Victoria A new economy.* https://bit.ly/42OT0KP.

[67] WA MainRoads. 2022. *Tonkin Gap project and associated work.* https://bit.ly/3XmZ7Dt.

[68] Tonkin Gap Alliance. n.d. *Project & Associated Works Sustainability Policy.* https://bit.ly/3yLjDUg

[69] Tonkin Gap Alliance. 2022. *Tonkin Gap Project and Associated Works: Annual Project Sustainability Report 2021-2022.* https://shorturl.at/cAI25.

[70] Waste Authority. 2012. *Waste avoidance and resource recovery strategy 2030: Western Australia's waste strategy.* https://bit.ly/3CFUj3o.

[71] Development WA. 2019. *Hamilton senior high school redevelopment.* https://bit.ly/3Xkjktw.

[72] TABEC. 2022. *115 Hamilton Hill Stage 1.* https://bit.ly/3Emy8zd

[73] PWEA/WALGA. 2016. *Specification for the supply of recycled road base.* https://bit.ly/3DndvCJ.

[74] Liu, H., Long, H. and Li, X. 2020. Identification of critical factors in construction and demolition waste recycling by the grey-DEMATEL approach: A Chinese perspective,' *Environmental Science and Pollution Research.* 27, pp.8507-8525. https://doi.org/10.1007/s11356-019-07498-5.

[75] Shooshtarian, S., Caldera, S., Maqsood, T., Ryley, T. and Khalfan, M. 2021. An investigation into challenges and opportunities in the Australian construction and

demolition waste management system. *Engineering, Construction and Architectural Management*, 29(10), pp.4313-4330. https://doi.org/10.1108/ECAM-05-2021-0439.

[76] Shooshtarian, S., Caldera, S., Maqsood, T., Ryley, T., Wong, P. S. P. and Zaman, A. 2022. Analysis of factors influencing the creation and stimulation of the Australian market for recycled construction and demolition waste products. *Sustainable Production and Consumption*, 34, p.163-176. https://doi.org/10.1016/j.spc.2022.09.005.

[77] Shooshtarian, S., Hosseini, M. R., Kocaturk, T., Arnel, T. and Garofano, N. T. 2023. Circular economy in the Australian AEC industry: investigation of barriers and enablers, *Building Research and Information*, 51(1):56-68. https://doi.org/10.1080/09613218.2022.2099788.

[78] Shooshtarian, S., Maqsood, T., Wong, P. S. P., Yang, R. J. and Khalfan, M. 2020. Review of waste strategy documents in Australia: analysis of strategies for construction and demolition waste,' *International Journal of Environmental Technology and Management*, 23(1), pp.1-21. http://dx.doi.org/10.1504/IJETM.2020.10030234.

[79] Wahi, N., Joseph, C., Tawie, R. and Ikau, R. 2016. Critical review on construction waste control practices: Legislative and waste management perspective,' *Procedia - Social and Behavioral Sciences*, 224, pp.276-283. https://doi.org/10.1016/j.sbspro.2016.05.460.

[80] Shooshtarian, S., Caldera, S., Ryley, T., Maqsood, T., Zaman, A. and Wong, P.S.P. 2022. *The role of education in the Australian circular built environment: Analysis of educational programs and construction and demolition waste management.'* The 55th International Conference of the Architectural Science Association 2022: How can Design Enhance the Quality of Life?, Perth, Australia. https://shorturl.at/kmwI3.

[81] Yu, B., Wang, J., Liao, Y., Wu, H. and Wong, A. B. 2021. Determinants affecting purchase willingness of contractors towards construction and demolition waste recycling products: an empirical study in Shenzhen, China. *International Journal of Environmental Research and Public Health*. 18(9), p.4412. https://doi.org/10.3390%2Fijerph18094412.

[82] Elena, S., Konstantinos, K. and Giorgos, G. 2022. Integrated management of construction and demolition waste as key factor of urban circular economy. *Journal of Sustainability and Environmental Management*. 1(2), pp.197-209. https://shorturl.at/ouwz4.

[83] Liu, X., Ren, H., Wu, Y. and Kong, D. 2013. An analysis of the demonstration projects for renewable energy application buildings in China. *Energy Policy*, 63, pp. 382-397. https://doi.org/10.1016/j.enpol.2013.08.091.

[84] Shooshtarian, S., Maqsood, T., Caldera, S. and Ryley, T. 2022. Transformation towards a circular economy in the Australian construction and demolition waste management System. *Sustainable Production and Consumption*. 30, pp. 89-106. https://doi.org/10.1016/j.spc.2021.11.032.

[85] Balador, Z. 2020. *Increasing the use of reclaimed and recycled building materials in New Zealand: Stakeholder perceptions*. PhD Thesis. Victoria University of Wellington. New Zealand. https://doi.org/10.26686/wgtn.17148512.v1.

[86] Udawatta, N. Zuo, J., Chiveralls, K., Yuan, H., Zillante, G. and Elmualim, A. 2018. Major factors impeding the implementation of waste management in Australian construction projects, *Journal of Green Building*, 13(3), pp.101-121. https://doi.org/10.3992/1943-4618.13.3.101.

[87] Tam, V. W., Kotrayothar, D., Loo, Y. C. 2009. On the prevailing construction waste recycling practices: A SouthEast Queensland study. *Waste Management and Research*. 27, 167–174. https://doi.org/10.1177/0734242x08091864.

[88] Active Sustainability. 2020. *WA Construction Resources - Recovered Construction and Demolition Materials Resource Guide*. https://bit.ly/36a9Po1

[89] Shooshtarian, S. and Maqsood. T. 2021. *How does product certification help construction and demolition waste resource recovery? The Fifth Estate*. https://bit.ly/3u0SNoc

[90] WSP Environmental and TRL Ltd. 2005. *Evaluating options for declaring recycled content in construction products and projects.* https://bit.ly/3OPK0NK

[91] Silva, R. V., De Brito, J. and Dhir, R. K., 2017. Availability and processing of recycled aggregates within the construction and demolition supply chain: A review. *Journal of Cleaner Production*, 143, pp.598-614. https://doi.org/10.1016/j.jclepro.2016.12.070.

[92] Gaffar, S. 2019. *How we can recycle more buildings*. The Conversation. https://bit.ly/3R1iyyu

[93] Su, Y., Si, H., Chen, J. and Wu, G., 2020. Promoting the sustainable development of the recycling market of construction and demolition waste: A stakeholder game perspective. *Journal of Cleaner Production*, 277, p.122281. https://doi.org/10.1016/j.jclepro.2020.122281.

[94] Waldmann, D. and Thapa, V. B., 2015. Use of recycled concrete in construction in Luxembourg. *Cahier Scientifique*-Revue Technique Luxembourgeoise. https://shorturl.at/auBO9.

[95] Wang, S., Tae, S., Kim, R. 2019. Development of a green building materials integrated platform based on materials and resources in G-SEED in South Korea. *Sustainability*, 11, 6532. https://doi.org/10.3390/su11236532.

[96] Equilibrium. 2019. *Review of standards and specifications for recycled content products*. Department of Agriculture, Water and the Environment. Melbourne, Australia. https://bit.ly/3HROgdz.

[97] Bao, Z. and Lu, W., 2020. Developing efficient circularity for construction and demolition waste management in fast emerging economies: Lessons learned from Shenzhen, China. *Science of the Total Environment*, 724, p.138264. https://doi.org/10.1016/j.scitotenv.2020.138264.

[98] Blue Environment. 2018. *National Waste Report.* Retrieved from https://bit.ly/3o1IBWy.

[99] Environmental Product Declaration. 2020. *Repurpose It - Recovered Mineral Aggregates*. https://bit.ly/3QPI3mr.

[100] GECA. 2023. *Sustainable solutions for planet and people*. https://bit.ly/3YuziRe.

[101] Holcim. 2023. *ViroDecs™ environmental product declaration.* https://bit.ly/3yiJWjk.

[102] GlobalGreenTag. 2022. *About Global GreenTag*. https://bit.ly/3XTsRHy.

[103] GBCA. 2015. *Assessment process for product certification schemes.* https://bit.ly/3wJAJQF.

[104] GBCA. 2023. *Responsible Products Framework.* https://bit.ly/3YvkVwe.

[105] Green Tag. 2016. *International Standard: Public Version 4.0.*

[106] ISO 14021. 2016. *Environmental labels and declarations — Self-declared environmental claims (Type II environmental labelling).* https://bit.ly/3XCmDMs.

[107] ISO 14024. 2018. *Environmental labels and declarations -Type I environmental labelling- Principles and procedures.* https://bit.ly/3HsdIre.

[108] DCEEW. 2022. *Nature Positive Plan: better for the environment, better for business*. https://bit.ly/41L77Aq.

[109] Miller, N. 2021. *The industry creating a third of the world's waste*. BBC, Future Planet: Sustainability. Published 16 December 2021. https://shorturl.at/zGPY7.

[110] SA EPA. 2022. *Container deposits. EPA South Australia*. Retrieved from https://shorturl.at/nzGP0.

[111] NSW EPA. 2019. *NSW Circular Economy Policy Statement Too Good To Waste.* Retrieved 3 December 2022, 2022, from https://shorturl.at/cdwFX.

[112] The State of Victoria Department of Environment, Land, Water and Planning, 2020. *Recycling Victoria a new economy. Retrieved 4 December 2022*, 2022, from https://shorturl.at/nCMTV.

[113] ACT Government. 2020. *ACT Waste Management Strategy Towards a sustainable Canberra 2011–2025 Reducing waste and recovering resources to achieve a sustainable, carbon-neutral Canberra*. https://shorturl.at/alyTY.

[114] The Waste Authority. 2019. *Waste Authority Annual Report 2018–19.* https://shorturl.at/movxN.

[115] Tasmanian Government. 2019. *Draft Waste Action Plan Consultation Draft.* https://shorturl.at/fGILN

[116] ACT Government. 2011. *ACT Waste Management Strategy Towards a sustainable Canberra 2011–2025 Reducing waste and recovering resources to achieve a sustainable, carbon-neutral Canberra*. https://shorturl.at/jprGP.

[117] Northern Territory Government. 2022. *Northern Territory CIRCULAR ECONOMY STRATEGY 2022 – 2027.* Retrieved 3 December 2022. https://shorturl.at/aeyLN.

[118] Queensland Government. 2019. *Waste Management and Resource Recovery Strategy.* https://shorturl.at/mnwI5.

[119] Green Industries SA. 2020. *SUPPORTING THE CIRCULAR ECONOMY SOUTH AUSTRALIA'S WASTE STRATEGY 2020-2025.* Retrieved 3 December 2022, 2022, from https://rb.gy/sgquo5.

[120] Skillington, K., Crawford R. H. 2020. *Design for Dematerialisation: an approach for reducing a building's embodied environmental flows.* Ghaffarianhoseini, A (Ed.) Ghaffarianhoseini, A (Ed.) Nasmith, N (Ed.) 54th International Conference of the Architectural Science Association 2020, 2020-November, pp. 355-364. The Architectural Science Association (ANZAScA). The 54th International Conference

of the Architectural Science Association (ANZAScA) 2020. Melbourne, Australia. https://rb.gy/elhoug.

[121] Ajayi, S. O., Oyedele, L. O., Bilal, M., Akinade, O. O. Alaka, H. A. Owolabi, H. A. and Kadiri, K. O. 2015. Waste effectiveness of the construction industry: Understanding the impediments and requisites for improvements. Resources, Conservation and Recycling, 102: 101-112. https://doi.org/10.1016/j.resconrec.2015.06.001.

[122] Shooshtarian, S., Hosseini, M. R., Kocaturk, T., Arnel, T. and Garofano, N. T. 2023. Circular economy in the Australian AEC industry: investigation of barriers and enablers. *Building Research and Information*. 51(1): 56-68. https://doi.org/10.1080/09613218.2022.2099788.

[123] Green Building Council of Australia (GBCA). 2021. *A circular economy discussion paper*. Retrieved 30 October 2022, from https://shorturl.at/cefY0.

[124] Guerra, B. C. and Leite, F. 2021. Circular economy in the construction industry: An overview of United States stakeholders' awareness, major challenges, and enablers. *Resources, Conservation and Recycling*. 170: 105617. https://doi.org/10.1016/j.resconrec.2021.105617.

[125] Ahn, Y. H., Pearce, A. R., Wang, Y. and Wang, G. 2013. Drivers and barriers of sustainable design and construction: The perception of green building experience. *International Journal of Sustainable Building Technology and Urban Development* 4(1): 35-45. https://doi.org/10.1080/2093761X.2012.759887.

[126] Shahid, M. U., Thaheem, M. J. and Arshad, H. 2022. Quantification and benchmarking of construction waste and its impact on cost – a case of Pakistan. *Engineering, Construction and Architectural Management*. 30(6): 2304-2333. https://doi.org/10.1108/ECAM-07-2019-0375.

[127] Akinade, O. O., Oyedele, L. O., Ajayi, S. O., Bilal, M., Alaka, H. A., Owolabi, H. A., Bello, S. A., Jaiyeoba, B. E. and Kadiri, K. O. 2017. Design for Deconstruction (DfD): Critical success factors for diverting end-of-life waste from landfills. *Waste Management* 60: 3-13. https://doi.org/10.1016/j.wasman.2016.08.017.

[128] Udawatta, N., Zuo, J., Chiveralls, K. and Zillante, G. 2015. Improving waste management in construction projects: An Australian study. *Resources, Conservation and Recycling* 101: 73-83. https://doi.org/10.1016/j.resconrec.2015.05.003.

[129] DGNB. 2020. *PRO1.4/Sustainability aspects in tender phase.* Retrieved 24 Semptember 2022, from https://shorturl.at/cmBL3.

[130] Carvalho Machado, R., Artur de Souza, H., and De Souza Veríssimo, G. 2018. Analysis of Guidelines and Identification of Characteristics Influencing the Deconstruction Potential of Buildings. *Sustainability* 10(8): 2604. https://doi.org/10.3390/su10082604.

[131] Dams, B., Maskell, D., Shea, A., Allen, S., Driesser, M., Kretschmann, T., Walker, P. and Emmitt, S. 2021. A circular construction evaluation framework to promote designing for disassembly and adaptability. *Journal of Cleaner Production* 316: 128122. https://doi.org/10.1016/j.jclepro.2021.128122.

[132] Crowther, P. 2018. A taxonomy of construction material reuse and recycling: designing for future disassembly. *European Journal of Sustainable Development* 7(3): 355-363. https://doi.org/10.14207/ejsd.2018.v7n3p355

[133] Eckelman, M. J., Brown, C., Troup, L., Wang, L. Z., Webster, M. D., and Hajjar, J. F. 2018. Life cycle energy and environmental benefits of novel design-for-deconstruction structural systems in steel buildings. *Building and Environment*. 143(1): 421-430. https://doi.org/10.1016/j.buildenv.2018.07.017.

[134] Çetin, S., Gruis, V. and Straub, A. 2022. Digitalization for a circular economy in the building industry: Multiple-case study of Dutch social housing organizations. *Resources, Conservation and Recycling Advances* 15, 200110. https://doi.org/10.1016/j.rcradv.2022.200110.

[135] Zoghi, M., Rostami, G., Khoshand, A. and Motalleb, F. 2022. Material selection in design for deconstruction using Kano model, fuzzy-AHP and TOPSIS methodology. *Waste Management and Research* 40(4): 410-419. https://doi.org/10.1177/0734242X211013904.

[136] ARUP. n.d. *Circular Buldings Toolkit.* Retrieved 25 September 2022, from https://shorturl.at/ewxyz.

[137] Clift, R., and Doig, A. 1995. *Developing Life Cycle Inventories for Waste Management*.

[138] Landells, E., Naweed, A., Pearson, D. H., Karunasena, G. G. and Oakden, S. 2022. Out of Sight, Out of Mind: Using Post-Kerbside Organics Treatment Systems to Engage Australian Communities with Pro-Environmental Household Food Waste Behaviours. *Sustainability*. 14(14). https://doi.org/10.3390/su14148699.

[139] Attia, S. 2018. *Regenerative and Positive Impact Architecture: Learning from Case Studies*. https://doi.org/10.1007/978-3-319-66718-8.

[140] DGNB. 2020. *ENV2.3 / Land use.* from https://shorturl.at/ikF69.

[141] Lemaitre, C., Braune, A., Ruiz Durán, C. and Geiselmann, D. 2019. *Fostering the Integration of Circular Economy Aspects into the Construction and Real Estate Industry.* IOP Conference Series: Earth and Environmental Science 290(1): 012021. http://dx.doi.org/10.1088/1755-1315/290/1/012021.

[142] Bertino, G., Kisser, J., Zeilinger, J., Langergraber, G., Fischer, T. and Österreicher, D. 2021. Fundamentals of building deconstruction as a circular economy strategy for the reuse of construction materials. *Applied Sciences* 11(3): 939. https://doi.org/10.3390/app11030939.

[143] Othman, A. A. E. and El-Saeidy, Y. A. 2022. Early supplier involvement framework for reducing construction waste during the design process. *Journal of Engineering, Design and Technology* ahead-of-print(ahead-of-print). https://doi.org/10.1108/JEDT-10-2021-0566.

[144] Yu, Y., Yazan, D. M., Bhochhibhoya, S. and Volker, L. 2021. Towards Circular Economy through Industrial Symbiosis in the Dutch construction industry: A case of recycled concrete aggregates. *Journal of Cleaner Production* 293: 126083. https://doi.org/10.1016/j.jclepro.2021.126083.

[145] Allwood, J. M. 2013. Squaring the Circular Economy The Role of Recycling within a Hierarchy of Material Management Strategies. The Netherlands. https://doi.org/10.1016/B978-0-12-396459-5.00030-1.

[146] Torgautov, B., Zhanabayev, A., Tleuken, A., Turkyilmaz, A., Mustafa, M. and Karaca, F. 2021. Circular economy: challenges and opportunities in the construction sector of Kazakhstan. *Buildings*. 11(11): 501. https://doi.org/10.3390/buildings11110501.

[147] Waster and Resources Action Programme. 2015. *Cutting embodied carbon in construction projects.* https://greenbuildingencyclopaedia.uk/wp-content/uploads/2015/11/WRAP-FINAL-PRO095-009-Embodied-Carbon-Annex.pdf.

[148] IPCC. 2018. *Global Warming of 1.5°C. Intergovernmental Panel on Climate Change*. Available at https://www.ipcc.ch/sr15/.

[149] United Nations. 2015. *Paris Agreement*. Available at https://shorturl.at/pxKNS

[150] UNEP. 2022. *2022 Global Status Report for Buildings and Construction: Towards a Zero-emission, Efficient and Resilient Buildings and Construction Sector*. Nairobi.

[151] Junnila, S., Horvath, A., and Guggemos, A. A. 2006. Life-cycle assessment of office buildings in Europe and the United States. *Journal of Infrastructure Systems*, 12(1), 10–17. https://doi.org/10.1061/(ASCE)1076-0342(2006)12:1(10).

[152] Ortiz, O., Castells, F., and Sonnemann, G. 2009. Sustainability in the construction industry: A review of recent developments based on LCA. *Construction and Building Materials*, 23(1), 28–39. https://doi.org/10.1016/j.conbuildmat.2007.11.012.

[153] Doe, J. and Smith, R. 2022. Design for deconstruction in circular economy: A review. *Journal of Sustainable Construction*. 1765: 105847.

[154] Charef, R., Morel, J. C., and Rakhshan, K. 2021. Barriers to implementing the circular economy in the construction industry: A critical review. *Sustainability*, 13(23), 12989. https://doi.org/10.3390/su132312989.

[155] EMF, 2020. *What is a circular economy?*. Available at https://shorturl.at/JOP14

[156] F[1]inch, G., Marriage, G., Pelosi, A., and Gjerde, M. 2021. Building envelope systems for the circular economy; Evaluation parameters, current performance and key challenges. *Sustainable Cities and Society*, 64, 102561. https://doi.org/10.1016/j.scs.2020.102561.

[157] Rios, F. C., and Grau, D. 2019. Circular economy in the built environment: Designing, deconstructing, and leasing reusable products. *Encyclopedia of Renewable and Sustainable Materials*. 5, 338-343. https://doi.org/10.1016/B978-0-12-803581-8.11494-8.

[158] Akhimien, N. G., Latif, E., and Hou, S. S. 2021. Application of circular economy principles in buildings: A systematic review. *Journal of Building Engineering*, 38, 102041. https://doi.org/10.1016/j.jobe.2020.102041.

[159] Akinade, O., Oyedele, L., Oyedele, A., Davila Delgado, J. M., Bilal, M., Akanbi, L., Ajayi, A. and Owolabi, H. 2020. Design for deconstruction using a circular economy approach: Barriers and strategies for improvement. *Production Planning and Control*, 31(10), 829-840. https://doi.org/10.1080/09537287.2019.1695006.

[160] Minunno, R., O'Grady, T., Morrison, G. M., Gruner, R. L., and Colling, M. 2018. Strategies for applying the circular economy to prefabricated buildings. *Buildings*, 8(9), 125. https://doi.org/10.3390/buildings8090125.

[161] Minunno, R., O'Grady, T., Morrison, G. M., and Gruner, R. L. 2020. Exploring environmental benefits of reuse and recycle practices: A circular economy case study of a modular building. *Resources, Conservation and Recycling*, 160, 104855. https://doi.org/10.1016/j.resconrec.2020.104855.

[162] Lange, R. and Morrison, G. 2019. Legacy Living Lab – visualisation of smart data and peer-to-peer trading of local renewable energy and alternative water systems from WGV and beyond. Low Cabron Living CRC, Perth, Australia. Available at https://apo.org.au/node/251506.

[163] Rahla, K. M., Mateus, R., and Bragança, L. 2021. Implementing circular economy strategies in buildings—from theory to practice. *Applied System Innovation*, 4(2), 26. https://doi.org/10.3390/asi4020026.

[164] O'Grady, T., Minunno, R., Chong, H. Y., and Morrison, G. M. 2021. Design for disassembly, deconstruction and resilience: A circular economy index for the built environment. *Resources, Conservation and Recycling*, 175, 105847. https://doi.org/10.1016/j.resconrec.2021.105847.

[165] National Waste Policy. 2018. *Less Waste. More Resources*, in Department of Agriculture, Fisheries and Forestry. Canberra, Australia. https://rb.gy/36b910.

[166] Senate Environment and Communications References Committee. 2018. *Never waste a crisis: the waste and recycling industry in Australia*, Parliament of Australia Canberra, Australia. https://rb.gy/wct2th.

[167] Crocker, R. and Lehmann, S. 2013. *Motivating change: Sustainable design and behaviour in the built environment*, ed., Routledge, London, UK.

[168] Chapman, J., Skinner, N. and Searle, S. 2013. *Working towards sustainability: exploring the workplace as a site for pro-environmental behavioural change*, In Motivating change: sustainable design and behaviour in the built environment, Routledge, London, UK, 88-102. https://doi.org/10.4324/9780203482087.

[169] Udawatta, N., Zuo, J., Chiveralls, K. and Zillante, G. 2015. Attitudinal and behavioural approaches to improving waste management on construction projects in Australia: benefits and limitations, *International Journal of Construction Management*, 15(2), 137-147. https://doi.org/10.1080/15623599.2015.1033815.

[170] Crock, M., Baker, J. and Turner-Walker, S. 2013. *Open Universities Australia: The evolution of open access and online education opportunities*, In Global challenges and perspectives in blended and distance learning, IGI Global, 83-98. https://rb.gy/yc2ske.

[171] Shooshtarian, S., Hosseini, M. R., Kocaturk, T., Ashraf, M., Arnel, T. and Doerfler, J. 2021. *The Circular Economy in the Australian Built Environment: The State of Play and a Research Agenda Prepared*, Deakin University. https://rb.gy/6fgdxw.

[172] GBCA. 2022. *Annual member survey ends 2021 on a high for GBCA*, Green Building Council of Australia, Sydney, Australia.

[173] Alchemy Living and Learning. 2014. *Report on professional development needs of waste and water industry trainers to educate for sustainability.* Available at: https://shorturl.at/bquCO.

[174] Younes, M. 2017. *The ADKAR Change Management Model.* https://shorturl.at/eqvZ8.

[175] Porsci. 2022. *The Prosci ADKAR Model.* Retrived from https://www.prosci.com/methodology/adkar

[176] Bushnell, D. S. 1990. 'Input, process, output: A model for evaluating training. *Training and Development Journal*, *44*(3): 41-44.

[177] Brinkerhoff, R. O. 1987. *Achieving results from training: How to evaluate human resource development to strengthen programs and increase impact.* John Wiley and Sons Inc. United States

[178] Kaufman, R., and Keller, J. M. 1994. Levels of evaluation: Beyond Kirkpatrick. *Human Resource Development Quarterly*, 5(4): 371-380. https://doi.org/10.1002/hrdq.3920050408.

[179] Meera. 2022. *My environmental education evaluation resource assistant*, University of Michigan, Michigan, US. https://meera.seas.umich.edu/.

Index

A

B

C

D

E

G

H

I

K

L

M

N

O

P

Q

R

S

T

U

V

W

Y

About the Authors

Professor Tayyab Maqsood has been involved in research related to C&D waste management for over a decade. He is leading the C&D waste management research group at RMIT and is the founder of Australia's first National Construction and Demolition Waste Research and Industry Portal.
He is a strong proponent of the circular economy or cradle-to-cradle approach. He is also Associate Dean of Project Management in the School of Property, Construction and Project Management at RMIT University.

Dr. Salman Shooshtarian holds a PhD in the Built Environment and is currently working as a Senior Lecturer at RMIT University. He is an active researcher in the fields of sustainability, circular economy, and C&D waste management. He has experience in designing and performing empirical research projects and has established collaboration with different research teams for several research projects. He is the lead author of a book titled Construction and Demolition waste management in Australia (2021). He is an active member of the Waste Management and Resource Recovery Association of Australia.

Professor Peter SP Wong is the Associate Dean of the School of PCPM. As of 2022, he has more than 100 publications and has been awarded over AUD $1.8 million in research grants. Professor Wong is a chartered construction manager and a chartered quantity surveyor with solid industry experience. His knowledge and experience particularly in organisational and operational management in construction provides a fundamental basis for effective contribution to research projects in relevant fields.

A/Professor Atiq Zaman is s currently working at Curtin University, Western Australia. He is also a researcher at the Curtin University Sustainability Policy Institute (CUSP) and actively advocates for zero waste and circular economy solutions. He is one of the founding co-directors of the Global South Nexus research cluster at DBE and has been working as the Curtin Node Leader for the Sustainable Community and Waste Hub.

Dr Savindi Caldera is a Lecturer at the University of the Sunshine Coast and an Adjunct Research Fellow at the Cities, Research Institute, Griffith University. As an expert in lean and green manufacturing, cleaner production and rapid capacity building, she has been working with public and private sector colleagues to enable resilient natural and built environments. She has published over 50 publications and has secured research funding of over AUD$2.1 million.

Prof Tim Ryley is Professor of Aviation, within the School of Engineering & Built Environment, Griffith University. His research expertise principally covers air transportation and environmental issues, but he has led a series of multi-disciplinary transport and engineering-based projects, including C&D waste management and demand responsive transport. Prof. Ryley has led or co-led research projects totalling over $5 million, typically funded by research councils and industry.